The Simmering Pot of Space-time

A teenager's guide to surviving science

Sumer Vaid

FROG BOOKS

First published in India 2012 by Frog Books
An imprint of Leadstart Publishing Pvt Ltd
1 Level, Trade Centre
Bandra Kurla Complex
Bandra (East) Mumbai 400 051 India
Telephone: +91-22-40700804
Fax: +91-22-40700800
Email: info@leadstartcorp.com
www.leadstartcorp.com / www.frogbooks.net

Sales Office:
Unit: 122 / Building B/2
First Floor, Near Wadala RTO
Wadala (East) Mumbai 400 037 India
Phone: +91-22-24046887

US Office:
Axis Corp, 7845 E Oakbrook Circle
Madison, WI 53717 USA

Copyright © Sumer Vaid

All rights reserved. No part of this publication may be reproduced, stored in or introduced into a retrieval system, or transmitted, in any form, or by any means (electronic, mechanical, photocopying, recording or otherwise) without the prior written permission of the publisher. Any person who does any unauthorised act in relation to this publication may be liable to criminal prosecution and civil claims for damages.

ISBN 978-93-81836-20-0

Book Editor: Priya Khanna
Design Editor: Mishta Roy

Typeset in Book Antiqua
Printed at Repro India Ltd, Mumbai

Price — India: Rs 95; Elsewhere: US $4

To,

Sankalp,
If only you were here, brother. We linger on without you.

Mom,
who is never baffled at the oddity and impracticality of my dreams;

Dad,
the compass of my life;

Dadu,
my personal knowledge bank;

Dadi,
for the ever-lasting warm blessings;

Misha,
for the (not-so) innocent entertainment;

A special thanks to Ms. Rajni Bakshi. This book is and will always be yours.

And lastly, to every adolescent who finds science boring.

About the Author

Sumer Vaid is a 16-year old science fanatic who lives in Mumbai. He has been interested in science ever since he can remember and has been pursuing his passion since he was 10 years old by reading books, watching TV shows and devouring any scientific material he could lay his hands upon. He writes fiction in his free time and shares his stories through the Internet. Sumer has made it his personal mission to shatter the stereotype of physics as posed by the Indian society; as that of a burdensome subject or a profitable business or a means of gaining admission to top schools. Sumer wishes to introduce the secret arena of jostling physics to the regular teen, and tell him or her that physics or science as a whole is anything but lethal. Sumer attends the Ecole Mondiale World School and his interests are majorly supported by his family and school. Regardless of his musings in this book, physics and mathematics remain his favorite subjects at school.

Acknowledgements

Authors are evil. They claim the rights of a book, slither into its pages, get stamped on its cover and spine, take credit for all the goods (don't forget the cash) and then smile smugly. What about the work of the hundred other people that have helped craft the book? Well, I am personally in no position to blame any author for espionage because I am of the same breed. My name is on the cover and spine of this book, it has also found its way inside the pages and yes, I can't say that this book belongs to someone else. I mean it's mine! I wrote it, after all!

Nonetheless, there are loads of people who have contributed extensively to this piece of work. They have been the pen, paper and ink of this effort, without which I might as well be another dreamer and no-doer. I would firstly like to thank my aunt, Ms. Rajni Bakshi, the wizard of this book. With a wave of her wand, my crude manuscript sparkled into one hell of a book. Having accommodated all my objections, claims, wishes, tantrums and thoughts, she still managed to get across her point through my egoistic array of filters, for which I am truly grateful. Most importantly, I would like to thank her for the inspiration. She was responsible for switching on the author that lay inside me by encouraging a compulsive passion towards the language within my soul. Thank you for all the amazing help, *bhua*.

Then come mom, dad and Misha. My father has been the compass of my life, and it is upon his directions that I understood my love for science and came to realize its potential. My mother has supported all my radical and outrageous dreams with a calm and composed face, which I think is one tough job to do. The constant encouragement from my parents has proved essential in the molding of this book, and without them, my manuscript might be rotting on an old dusty shelf. Misha, my little sister, was always around for giving me a dose of the standard teenager's view of science. I couldn't have ever written in this perspective without her constant inputs. As a whole, my family has played an integral role in this book. I am proud to say that this book belongs more to the latter part of my name.

My grandparents have played a truly amazing role in getting this book where it is. Without my grandfather's infinitesimal knowledge and his compulsive urge to share this knowledge, how would my own quest for learning have begun? Without my grandmother's generous blessings, how would I have made it this far? Thank you *dadu, dadi, naani and badi daadi* for your love and goodwill.

I would like to thank my school, Ecole Mondiale World School, for giving me an ideal environment to hone my literary and scientific skill. In particular, I would like to thank Mrs. Shalini Ghosh, the physics teacher with whom I have had conversations about some of the most obscure points of Physics. She was always up for a talk; be it the fascinating concept of a multi-verse or the jostling arena of electrons and quarks, Mrs. Ghosh never faltered to argue and amaze me with the intricacies of science. She also offered me support during all times and was a true critic of the book. Furthermore, I would also like to thank my Mathematics teacher, Mr. Chiryanth Francis, for keeping my love for mathematics blazing at all times. Regardless of what I've written in this book, Math still remains my favorite subject, sir! Finally, I would also like to thank all my friends for sticking around when I was too busy working on this book and couldn't even spare a minute for them. You all are and will always remain the ink of this book.

I would like to particularly thank Dhruva Shetty, my younger cousin and the first reader of the book who after having devoured a few paragraphs of the crude manuscript thought that this book had the potential to win an award. Thank you for your encouragement and support, brother!

I would also like to thank Jatin Mishra for his critical view and his constant (almost irritating) encouragement. Without his constant bugging and reminders, this book would have never reached this far.

Lastly, I would like to thank God for giving me all that was essential to pen this book. I would like to thank my brave publishers, Leadstart, for giving a young lad a platform to communicate with the world; and sharing the risk of failure with a first-time author.

I take this opportunity in thanking anyone whom I have left out, and without further ado, urge you to flip the page.

Contents

Author's Note   11

Chapter I: Philosphysics   13
- *Holograms and Projections*
- *The Goldilocks Zone*
- *Las Vegas Luck*

Chapter II: Evil Scientists and Evil Dimensions   22
- *Dimension Folklore*
- Dead Babies and Dimensions
- Higher Dimensions

Chapter III: The Future of Physics   30
- The Invisibility Cloak
- Teleportation
- Time Travel

Chapter IV: The Smelly Jelly   51
- Why the brain?
- The Butter Brain Delicacy.
- Upgrading the Jelly

Are we just Holograms?

Why does the Goldilocks Zone make life possible?

How does living at different floors affect how slow or fast you age?

Where does infinity end?

Author's Note

It happened on one of my weekly visits to the Crossword bookstore. Strolling in the Science section and surrounded by a plethora of books, I sighted a text squatting in a nook of sorts. It was a medium-sized book, clasped between two mammoth-sized encyclopedias, almost gasping for attention. *Physics of the Impossible*, the cover read. I dusted off the front and opened the book. After 10 minutes of sincere observation and quick glances at the chapters, I was convinced that reading this book would be a life-changing experience.

A year later, after having swallowed several texts on Theories of Relativity and any physics-related material I could lay my hands on, I was lost in the realm of physics. Never have I seen the world in the same perspective after having read that book. Suddenly physics was all around, secretly whispering the most unbelievable things to me, about the universe, space and time and many more marvelous things. Physics had carved a secret universe of its own, that existed with supreme ignorance amidst our very own, explored only by the likes of Einstein and Newton, and I made it my personal goal to expose this vibrant universe to everyone who thought science was boring. I wanted to write a book.

A book that displayed the colorful spectrum of physics, how it stretched on endlessly, containing marveling facts and phenomenon and how it never failed to amaze. My mission was to delicately serve advanced physics upon a platter of daily-life examples and situations. As three years have passed from the time I picked up that text, I write with the intent to not only discover but to also communicate. To tell that tired high-school student that physics is beyond boredom. To tell that businessman that physics is profitable.

To tell parents that physics is beyond grades. Most of all, to tell the world that physics is marvelous. It is beyond boards filled with equations. It is beyond formulas, or theorems. It is much more than that.

For a start, how would you react if I told you that the key to living longer is not about fancy anti-ageing creams but about living on the ground floor? A rather absurd idea you would think, but it's scientifically identified! In fact, Einstein himself established the physics behind this idea.

What if I were to tell you that a massive bunch of bananas and a billion microwaves would be a good replacement for the sun? Hilarious, I know, but true nonetheless.

What if I tell you that while reading this you are sitting inside the valley of an invisible fabric, and you are also making several other valleys around you? Unbelievable, until I introduce you to Einstein's magic.

Would you believe the fact that you have the probability to disappear from here and materialize on Mars? A rather outrageous declaration you would say, but don't keep the book down just yet! Because all the claims are true, and as you read on you will find that physics is behind all of these uncanny statements.

So strap on your seat-belt and get ready to ride the rollercoaster of physics!

Chapter I
Philosphysics

It is awesome to travel to another solar system and see a bunch of blue tribes living in their fantastical world filled with glowing plants, floating mountains and dragon-like creatures. It's even cooler to strap on massive body armor and compound your body strength by ten and then run around in your neighbor's backyard, yanking all those expensive cars. Science fiction is cool. I mean, after all, sci-fi cooks up some fantastic movies and books for us. It's the stuff that takes us beyond the limits of our earthly homes into the dark void of the universe.

But what about plain old science outside fancy fiction? Is that boring physics crouching behind the science in science fiction cool too? Forget "cool", most of us frown and shed a tear when big daunting adults slowly utter the words "SCIENCE", which is often followed by something even more patronizing, like "HOMEWORK". But that's understandable. We never learn the cool science, do we? We have to make do with the boring old yaps that the apple-struck Newton discovered while he dozed under a tree, in a grand green orchard. I mean, even the damn guy who finally discovered gravity decided to give up on the wretched concept for a while and take a nap. And that friggin' apple had to fall on his head? The apple too, mind you, had to be light enough so it only gently caressed Newton's upper head. It needed just enough of a push to get him out of his sleep and ask "Why the hell did the apple fall down?" Had the apple been lighter, Newton would never have woken up and our textbooks would have been thinner. Had the apple been heavier, Newton would hopefully have had a memory-loss and the perils of calculus and gravitational physics would have been lost forever. But all those things don't happen in the real world. In the real world, everything from the weight of the apple to the position of a scientist's head has to be perfect. Welcome to the nonsensical world of scientific discovery.

Humans are (apart from the lovely entertainment they provide) quite useless and dumb. We know as much about our surroundings and environment perhaps, as an average third grader knows about differential calculus. Forget our surroundings; let's talk about our very own body. We know as much about our *own* human brain than Da Vinci knew about Michael Jackson (which he quite obviously did not, so you get the idea).

Scaling back up, we don't know what more than 70% of our universe is made up of. Around 30% of the universe appears to be made up of the extremely familiar planets, galaxies and blah. But all the other 70% seems to be made up of a weird substance known as dark matter, something that we don't know a whiff about. Back again on Earth, things are worse. Hell, we didn't even have confirmed evidence that there is a mantle beneath our feet. Weird isn't it? You are standing or sitting on something firm as you read this book, and yet we have no exact idea or evidence of what is beneath our happy little crust.

Even then, mind you, our textbooks seem ignorant to dieting, filling themselves with pages and pages of apparent facts and information. But the sad truth of this pit of knowledge is that most of this apparent knowledge that humans cherish is *theory*.

Which means there's a fat chance that we wake up one fine morning and find ourselves confronted with massive levitating slabs of iron saying "Game Over". Before we even know it, a higher-dimensional alien has switched off our universe because he was tired of playing the same old video game. In other words, we aren't even completely sure that we *exist*. So how can we be so sure that the universe we perceive is in fact reality?

We may as well be a hologram created by distant creatures for a laugh or joke. I know this sounds a bit like claiming Einstein's coolness over Charlie Sheen, but hang on. Let's see why our existence is so unlikely in this bubble of our universe.

What about Life: Las Vegas Luck or an Evil Alien Plan?

Scientists believe that the way life was kicked off on the third rock from the sun was done in circumstances that were a bit too perfect. Think of it like a recipe; as if you are preparing a chocolate cake. The only difference is, even the most microscopic mistake could blast your wretched cake into a million pieces. Everything, from every grain of flour you pour into the baking dish, to the last drop of the egg yolk you fling in, has to be measured and exact and according to the recipe. Scientists think that these were the conditions entrusted upon Earth's chef. But then, that's where the damn questions really come up. How the hell did the chef manage to prepare a perfect rock with perfect holes to pour salty water in and perfect flat lands to grow trees on, amid the simmering temperatures of the universe and the battleground of flinging meteor missiles and radiation bullets? Besides this, questions regarding the chef also arise. Who, in Christ's name, was the chef anyway? (Yes, there was a pun somewhere in there). And why didn't the other planets, such as Mars, not get cooked and fried enough to support life?

The answer to that is -- Earth got lucky. Mars got swamped. Earth won the million-dollar hand with the perfect chef and the perfect recipe, while Mars had to make do with red sand. Or at least, that's what scientists have been thinking for thousands of years. And our thick, bound textbooks have been stuffed with assumptions that Earth was lucky enough, (which is very very *very* lucky) to make do with life-forms rather than just chunks of red rocks. But is this really true?

Welcome to the (god-for-saken-are-you-kidding-me) Goldilocks Zone

We still can't figure out if we were really lucky enough to live but we do one thing for certain: the apparent 21[st]century scientists, (also known as average-sized bearded humans responsible for feeding our adolescent years with the likes of physics) have a weird fetish for fairytales. Why? 'Cause the only known place in the entire, whimsically amazingly large universe that can say "Oh yeah, I got the stuff to life you up", has been named after a bedtime story about a tiny blond girl sleeping in a bear's forest house. Don't know what I'm talking about? *Goldilocks*. That's what the scientists call the area of the solar system where life can be sustained. Outrageous, if you think of it. Well at least there's one thing in the universe that is not named after an ancient semi-nude Greek God.

So, the scientists decided to rid their normal custom of naming every scientific phenomenon in ancient Latin? No, not really. Scientists named this essential zone the Goldilocks Zone because of the perfection possessed by the zone. Remember the story of that bored blond girl in the forest? She always wanted everything to be perfect; neither too hot nor too cold (the porridge). Her bed was neither too big nor too small. Following the same ideology, life's demand for perfection matches Goldilocks' obsession for perfection with the perfect environment that the zone has created for us. Inside the Goldilocks Zone, it is neither too hot, nor too cold. The sun's rays are just right and hardly any cosmic radiation capable of frying makes it through the boundaries of the zone. In short, the zone gives us life.

Anyway, what makes the Goldilocks Zone so special? The specialness of the zone makes sure that you are sitting calmly, listening to Coldplay in the background and reading this book instead of watching your blood boil to a few thousand degrees or losing your pupils to meteor dust, or experiencing any other evils of the universe. That's what's so special about it. The Zone makes sure that we live in our bubble planet and stay happy without bursting into a million pieces. It makes sure that the sun's power is just enough to make the flowers bloom. It makes sure that a couple of gases stay glued onto the sky above our heads so we don't cough our lungs out. It makes sure that it is warm everyday instead of the millennia-dropping sub-zero temperatures right outside the Earth's atmosphere. That's what the Zone does. It makes sure we live in perennial joy while the rest of the universe out there tries to wage war. And the Zone is still named after a damn fairytale. Ironic. The key to the amazing things that the Zone does lies in its placement. The Goldilocks Zone is so life-able because it has been placed at an ideal distance from the Sun; to ensure that the Earth doesn't become an oversized crusted meatball when things get hot. Rather, the Goldilocks Zone is not even a materialistic existence. It has been simply defined (so rare in physics, eh?) as an area in the universe where all conditions are perfect for the formation of life – as we know it so far!

But this Goldilocks Zone thingy does not answer our question. We know now that the Earth is inside a special danger proof bag that helps the people on it live their life. We know that this special Zone relies on the precise placement of the Earth from the Sun. But then more questions come in. Is it really true that this entire Goldilocks Zone was created by "luck"? Or did someone in the universe decide that Earth would have life and shoved a rock in the Goldilocks Zone? And if so, then who in god's name is that person? These are some of the things that baffle even the most bearded geniuses. Let's see how far we have gotten into discovering who gave us this lucky break.

Physics is odd. It is weird. It does not agree with itself. It's a radical science. Things can vanish and re-appear. Lifeless electrons can be "conscious". Physics on a grand scale is unbelievable. The same physics that gives you pains and the perils of no sleep can be twisted to create the most senseless yet relevant theorems about the universe. One such theorem is called the "Holographic" theorem. In very simple words, some in the scientific community think that our universe is a fake three-dimensional projection. Very normal, isn't it? After all, I was tired of the reality anyway. Moving on, these scientists also think that the universe is a giant ball with everything inside it being a projection of sorts. Why do scientists think we are a hologram? Well, you can blame the flavorful quarks or the lifeless electrons for that.

Electrons are weird. What we have been told in school about electrons is true to some extent, but then again, it's a frail theory. We have been told electrons are negative particles that whiz around nuclei. We have been told that electrons make reactions happen, but more importantly, they satisfy our greedy need for power by pumping out watts of electricity that do everything, from running your laptop to cooling your room. But the true nature of electrons goes far beyond all of this. And this true nature has left those Harvard graduates very confused. How?

Let's start with one electron. We take that electron and another different electron. We tune them together to a similar frequency (just like your FM), so they both are rotating in a clockwise direction. We take the two electrons, put one on Mars and one on Earth. We make sure that both the electrons are spinning.

Now, here is where the jaw-dropping, heart stopping, mind breaking fact comes in. If I make any change to any one of the electrons, the other electron would *automatically* and *instantaneously* change so that if follows the exact property of the other changed electron. And the most stunning fact about this is that the distance between those two electrons does not matter one bit. In very simple words, scientists at CERN and Fermilab have found out that two electrons running on the same frequency can exchange data at a speed faster than that of light, which is like saying that the last 100 years' worth of physics is as pointless as a doormat in a desert. Why? 'Cause Einstein's genius says that nothing, absolutely nothing with a measurable mass in the universe can transmit information or travel faster than the speed of light. That is the ultimate law of the universe. It is THE 'studly' law that scientists have been nodding their heads to from the last century or so. Almost every new discovery in Physics is somehow supported by that Einsteinian law.

So the shattering of that law would ensure the shattering of all modern physics as we know it. Our textbooks will finally diet, information will be less, and teachers won't have to crib about finishing their corrections in time.

But we forget, not all scientists are bearded and humans. Some physicists are bearded and beasts. And the beasts are the ones who prove all these bizarre theorems. So, along came Hogan and a bunch of other physicists claiming the universe was a hologram. Let's take an easy example to understand what these geniuses said.

The Simmering Pot of Space-time

We are surrounded. Like a fish in a pot of water is not aware that it dwells in water, we are unconsciously living in a pot of a different material, known as "Space-time". Or at least, Einstein says so. According to him, the entire universe is a like a pot which is filled to the brim with an intangible substance known as Space-time. Space-time, Einstein claims, is the key to our universe. It surrounds every star, every planet, every galaxy and every chewing gum.

So, if everything is submerged in the same liquid then I can push one object here and get another object, miles away, to move, right? Eh…How?! Through waves, duh! If I make a gigantic wave in this space-time, I will end up moving planets, stars and the entire solar-system; just like the waves in an ocean can move strewn bottles and garbage through sheer wave force. But that's where the problem comes in.

Now, imagine the same fish in a pot of water, to be inside an entire gigantic ocean. The fish is less than a tiny speck in the entire ocean, and even with all its power, it won't be able to make a wave big enough to move entire ships and boars. That's the case with us too. In fact, humans are so microscopic when compared to the universe that we can't even create miniature waves in the space-time that is around us. We can cause microscopic changes in this material, but it might take several million years before we have enough power to actually make noticeable changes in this material.

But, my friends, we are not all that dumb. Even though we can't make changes in this material, we can definitely make machines that detect changes in this material. So along comes a monster of a machine known as GEO600, to detect even the most microscopic changes in Space-time. Think of it like a crazy goldfish scientist, building something to detect small ripples in its small pot of water. The machine took several years and several hundred million dollars to build. So what happened when scientists decided to yell, "To infinity and beyond"?

When the mammoth machine was switched on, all we heard was

a constant buzzing. That's it. Like someone had tuned a radio between two stations and left it there.

Then came the bad part; the scientists lost it. They tore off their beards, burnt their overcoats and buried their spectacles in dismay.

But then along came their savior in a thousand dollar Armani overcoat, Craig Hogan.

"Hello!" he said. "Don't tear your damn beards off. Shave them with razors! And once you're done, don't forget to wash your hands! Then y'all hommies can listen to me. All that this pretty little noise tells us is that our pretty little Space-time fabric is not all that smooth as Einstein had said. It's just very rough! Why is it rough? 'Cause we're a damn hologram!

Think of it like a pot filled with dirty, mucky water. The scientist goldfish we earlier talked about, would study the dirty water inside the pot and find that the material that it was living in was rough and uneven. It would appear to the goldfish that the material it was living in was rough and strewn with weird particles (dirt). But the goldfish won't be able to feel these microscopic particles in its day-to-day swimming activities, as these particles are 'wayyy' too small to be felt by the fish. So, the goldfish scientist would have thought that (if it knew it was living inside a particular substance) the substance it was living in was very smooth. But upon scientific observation, the fish would go ballistic trying to figure out why the substance it was living in was so damn rough and weird; just like humans, who are currently tearing their hair out wondering if our universe is really a distant projection.

When you a have certain image and you size it very high, what happens? The picture becomes blocky and pixelates. The quality of the picture degrades and worsens and you can see the pixels that make the picture up. Same thing with the universe. When we zoom down into the hologram, we see roughness because we have gone too far in or zoomed too far down. And this leads us to believe that Space-time is nothing a but a distant projection and we are nothing but a massive hologram. If the entire universe was nothing but a giant image, and we went too far in, we would eventually see big blocks and pixels. And so said Craig Hogan, the universe is a damn giant hologram.

And that's pretty much all there is to it. Half the geniuses think we're a hologram. The other half thinks we ain't. But the point is, there is a fat chance that we are a hologram; which means we don't even exist really, but are just some sort of a virtual imagination. Pretty cool, huh? Like someone might just accidentally turn the projector's switch off and BAM. There goes our entire universe with your long held PS3 FIFA profile. But yeah, that's pretty much all I got for you now.

So now that we arrive onto questioning our existence, and claiming ourselves to be pretty little holograms, let's take a deeper look into the universe. Let's see what dimensions are and how they affect the probability of us being a hologram. That should give us a better picture about the universe - about why we're here in this universe and probably why math is such a boring subject.

Chapter II
Evil Scientists and Evil Dimensions

As if three dimensions weren't enough to summon a million perplexing formulas in our mathematic textbooks, some bearded geniuses have declared that the world is now made up of ten or eleven dimensions. We all know the common three dimensions; the ones that have ensured our high-school life can never get enough of volume, perimeter or area. They have also been christened by some rogue sophomores as the unholy trinity. Don't get it yet? Let me introduce you to the proud and the smugly, Length, Breadth and Height. Along the shores of the trinket universe, which is observed by our several meter long, upward-looking billion dollar optical beasts, mankind has never found anything that doesn't possess these three dimensions. Everything tangible that we have touched or observed has a height, width and breadth. So what happens when you have a new dimension?

Dimensions, like most of the other things present in our school textbooks, are a celestial pain. They seem to fit together with each other nicely, explaining all the physics to us and smiling down at scientists. But then their evil doppelgangers come slithering around and choke out a new dimension; which uproots the most of physics we have known in the last couple of centuries; the physics on which thick textbooks have been bound and forced into a junior's backpack and on which countless other important things have been done.

Dimension Folklore: Slade's Delusions

Something like this happened when a nutcase American wizard, Mr. Slade, dawdled down to London and started performing a late eighties version of Chris Mindfreak. He started intertwining rings without breaking or melting them, took coins out of coin filled bottles, without shattering or unscrewing them and other such holy-crap-how-did-you-do-that things; you get the idea. Well, sooner or later, he was sued and the court dragged him into custody, trying him for misleading his audience to believe bizarre illusions through some unknown illegal method (palmistry, as the British court defined it). To defend his notorious acts and his relentless magic, Slade puffed out his chest and said that he did his sorcery using a different fourth dimension, in which he had befriended "spirits". Appalled, the court planned to slam him with an even more vigorous sentence, when suddenly, out of nowhere, several revolutionary scientists jumped to his rescue. These scientists would go on to become Noble laureates, and were even at a time before their achievements, considerably respectable men.

Basically, if one of these scientists rose and spoke complete utter rubbish to an audience of 10, at least nine of those myopic fools would nod a dozen times at every word. So naturally, the court was eager to face the defendant's custody.

A particular important worldclass physicist by the name of Johann Zollner played a leading role in trying to save the sorry Slade's course in British prison. Zollner explained in layman's terms what Slade was trying to justify, and how his magic worked. Now, back in the early 1900's, a layman's term was not as a layman's term in the 21st century. So let me explain what good ol' Zollner claimed, in my own way.

Dead baby jokes are amusing and laughable as long as they are *jokes*. To evil scientists however, this is different. If these evil scientists had it their way, they would yank a baby off his cradle and drop him near a valley in the Grand Canyon. They would watch the plump little blob of flesh crawl down to the edge of a 50,000 feet drop to a valley, observe him looking down at the abominable abyss of darkness, gurgle something in an unknown dialect of baby language before watching him retreat back. "You slimy little porcupines, what did you think, huh? I would just fall off into a deep pit of crap?" the baby would yell. To scientists, however, this would seem nothing more than a mixture of goo's and gaa's. "The baby is hungry, feed him, Fred", one genius would probably say. But the point here that really is worth observation of the baby's action is not the fact that humans don't know baby language (we've known that ever since).

The fact that is worth noticing is that a newly born baby, who has never gone to school or spoken to anyone, knew that if he fell down the canyon, he would die. This experiment has been tested by scientists previously, with the same results showing up (not in such *harsh* conditions however, as the United Nations is still against Dead Baby experiments). So how in Merlin's unshaven beard, does a twaddle of a toddler know that we cannot *fall up* and that valleys are always *deep* but not *high*; without ever having been told so? Simple answer, say scientists. Our brain has been wired to think in three dimensions. Evolution has meddled with our brains over the eons to only perceive the world in three dimensions. Quite frankly, this has happened because evolution has only given a twat about survival.

Thousands of years ago, in a world where sabre-tooth cats often came hopping along, asking for sugar; quick instincts were necessary. And with our small, yet extremely hungry brain, we only developed our ability to react fast and think quickly amidst our confined world of length, width and breadth. This choice was much better than to gape around and perceive things in different dimensions to only get snacked on by a hairy beast. So, scientists say, when a baby is born, he automatically knows the normal three dimensions. Up, left and right, as he is told by a bunch of floating criss-crossed proteins known as genes, who was made to tell him so by Mr. Evolution. This is one of the things that good ol' Zollner said to defend the crackpot American Slade. According to him, we were hardwired to think in three dimensions; due to an unavoidable natural choice. And so, he said, what if there was another murky dimension lurking between these three? What if there was more to the world than length, height and breadth? What if there was a fourth dimension that we cannot think of or visualize because our brain doesn't have enough RAM to do so? What if Slade was actually right? To understand now, what Zollner said, let's think of something else.

Understanding Dimensions

We all know that people/objects and environments in paintings, laptop screens or any other intangible yet viewable world are in two dimensions. How do we know this? Well, if you tried to grab at your Facebook profile in the hope to change your display picture manually, you would either end up with a broken finger or a scratched monitor screen (the latter being more likely). So, you ask, why can't we *reach out* at such two dimensional things? The answer is quite simple. Cause they're *two* dimensional.

There is no height in two dimensional worlds. Only two dimensions, length and breadth are present in these objects. Now imagine a world, with ONLY two dimensions. There is no height, so other people in this unlucky world can't "reach-out" upwards or downwards, but only towards their right and left.

If you were gaping down at this world, it would appear nothing different from a flimsy sheet of paper with crawling stickmen. The physics in this sheet of paper will prevent these stickmen from pointing towards you, as you watch them on the sheet of paper; because for this to happen, the stickman would have to come out of the stick-land; an event that is less likely to happen than for you to see a unicorn in your backyard. As there is no concept of height in this world, the evolution of this mystical land has made sure that these people can only perceive two dimensions, just like we can only see three dimensions.

Now think. What would happen if you grabbed a three dimensional ball (a normal ball in our normal world) and threw it into the two dimensional stickman world? You have three options to answer this question:

a) The ball would go right into the two-dimensional world and knock up an excuse for a quick game of basketball among the two dimensional stick people.

b) The ball would go into the two dimensional world, collide with a two dimensional dude and yank off his two dimensional head.

c) The ball would bounce back and smack you in the face instead of going into a fictional world filled with heightless small people.

No, you're not a pathetic primate who has a brain the size of a peanut if you thought that the answer to this question lay in the obviousness of a) or b). The ugly truth, however, is that the answer is c). Now, if you've got this right and a smug grin is appearing on your face, be warned. You aren't much of a genius either, because you tried out the damn thing by throwing a ball on a piece of flat paper (that's pretty smart, actually).

Or, you may just be wise enough to think more than the folk who thought they could throw stuff in two dimensional worlds. If doing such a thing was indeed possible, you could reach out into your laptop and pull out Windows and Apples. But we can't do such things because we are in a three dimensional world. This same rule applies to the two dimensional world as well. Trying to throw a three dimensional object into a two dimensional world is like trying to stuff an egg down a postbox with a flip of a rectangular opening (no, don't even think of boiling the damn thing). The egg would just not *fit* into the rectangular opening.

So, now we know some very important things that good ol' Zollner said.

- Everything tangible can be described accurately with three numbers, AKA the unholy trinity or Length, Breadth and Height.

- But, we would really be ignorant idiots to think that only what we feel, touch or see is actually present in our universe. For a goldfish, its spherical glass fish bowl is its entire universe. (A dead fish experiment proved it, so there's definitely no argument). The goldfish can't gape outside the fish bowl, so it presumes its cheap glass sphere is its entire universe and spends most of its life floating hopelessly in it. The fish isn't aware that it is being kept as a petty pet by a bunch of evolving primates who claim themselves to be the most intelligent species on Earth.

- So, these hidden *things* that may be present in our universe, therefore, can't be seen by humans either because they contain an extra dimension, something more than the unholy trinity that humans aren't hardwired to understand or perceive. (Don't start sweating just as yet, because even if there are other dimensions and we somehow manage to find them through scientific proof, the mathematics of that dimension won't snarl their way into our textbooks, exams and physics lessons for the next couple of decades or more. But, yes, we are still stuck with the unholy trinity).

Think of the whole humans-are-too-dumb-to-understand-more-than-three-dimensions situation like our two dimensional, heightless bored little people understand it. If you manage to squeeze into a two dimensional world and tell a particular shorty, "Hey, bro you've grown taller since I last saw that face of yours", that 2D guy would simply say, "Yeah, I have become lengthier because of em' swishin' hoops", and if you somehow managed to control your urge to laugh at this particular moment; you would understand the key to unlock the jail broken lock of two dimensions. The 2D dude is mentally incapable of understanding a third dimension, so to him, becoming taller is just becoming "lengthier". A 2D guy only grows in two directions, upwards and sidewards; unlike humans, who grow upwards (taller), sidewards (broader) and forwards (fatter).

So, after having learnt these rather weird claims that good ol' Zollner made, let's see what he said next. According to Zollner, the third dimension was literally *high* above the two dimensional plane. To Zollner, the words you are currently reading are in a two dimensional world, while you yourself are in a three dimensional world. So where exactly would we find the fourth dimension? *Above* the third dimension? But doesn't the third dimension already have an *"above"?* One can, after all, *reach up towards the sky!* But in the two dimensional world, those two dimensional shorties would also claim to reach *up.*

But their *up* or so called *height* in the two dimensional world, is nothing but a length to us, Similarly, a four-dimensional dude looking at us claiming that *"we can reach up",* would merely see the distance we reach upwards with as a common length of the fourth dimensional world. Think about it this way. To a 2D dude, a ball is just a mere circle. If you tried throwing a three dimensional ball or a sphere into a two dimensional world, it would simply bounce back. So, how exactly do you think a four dimensional ball would look like? Would the four dimensional ball actually even enter into our dimension? No. To a 4D dude, *it is us who appear to be inside a sheet of paper.* Like we perceive the two dimensional people as those "inside" a sheet of paper, or inside a plane, the 4D creatures would perceive us as the 2D people, as "flat" people in a "flat" world. So the four dimensional ball would just bounce back into the four dimensional world and break the thrower's nose, just like it did in the three dimensional plane.

Four dimensional objects are very exciting. Imagine going to school, entering the school premises, giving your attendance and then simply *vanishing* and appearing at the end of the day to go back home. If I went and literally *picked up* a two dimensional person from a two dimensional world, to the people around him, it would look like that person has completely vanished out of their world. They would go wild and start poking the air with crazy two dimensional spears, probably thinking that invisible monsters are plaguing their worthless two dimensional lands. It would be hilarious. Now imagine, what would happen if I take a coin out of the two dimensional world, I flip it and then put it back into the 2D world? Let's assume that this coin has a "P" on one side and an "H" on the other.

A coin in the two dimensional world however, can only be viewed as or resemble a circle; because a coin is really a cylinder in the three dimensional world. So this coin in the 2D world is nothing but a circle with a "P" written on it. Now, I take this coin out, rotate it in my three dimensional world, and put it back into the two dimensional world. As the two dimensional people can only view one side of the coin, they will be startled to find the 'P' gone and an 'H' present out of nowhere. They would not be able to flip the coin because a coin in their land is just a circle, not a cylinder. Similarly, what would happen if I took out an entire two dimensional person, rotated him, and then put him back in? The two dimensional people would go nuts trying to figure out why they can only view the rear of the two dimensional person's body, and would wonder where the front of his face, legs and torso have vanished to.

To finish all this off, Zollner said that three dimensional objects in the fourth dimensional world would be nothing but a scissor-cut shape. To us, in our 2D world, this would appear as a cylindrical bottle inside which a cylindrical coin is placed. Once this very same thing is taken into the fourth dimension, however, it will appear as if the bottle has been cut out of a sheet of paper with a cut-out circle at its center (the coin). But this will only happen to a 4D dude looking at the bottle and the coin. Therefore, because now the coin is literally "on-top" of the bottle instead "inside" it, the 4D dude can simply pick up the coin and put both the things back in our world. To us, this would appear as though the coin has come out of the bottle without the bottle being broken or unscrewed.

And that my friends, is the lovely secret of dimensions. Scientists hope to find all these other dimensions sometime in the future, and prove Zollner right. And dimensions play a major role in our eventual survival in a dying universe. But before we consider our death, let's see how far we have come into turning fiction into reality. Let's see what all we will attain and invent before we are faced with a dying existence. Let's see the Physics of the Future.

Chapter III
The Physics of The Future

The future is here (well, no shit genius). Humans are much more 'star-trek' developed than what most lads around here think they are. It's just that a lot of the kickass inventions that can truly change our life have already been theorized and already tested at a small scale; but the ultimate perfection required to suit humans is what is mostly missing. We have technologies that can teleport. We have technologies that can turn us invisible. We have technologies that can lead to a doctor dancing on "Mama Mia". We have the technologies that can make Star Trek's universe appear like a petty fairytale. But what's missing amid this plethora of awesomeness is the testing of these inventions on humans.

Essentially, we are far more developed than most of us know or thought. Very few people know that such sort of technologies ALREADY exist in our world. We can already teleport, become invisible and time travel on a small scale. Yesterday's fiction is today's reality, literally.

But today's awesome reality of teleportation and invisibility is only tested by the math wizards and physics gurus. Commoners like us don't get a chance to get our hands onto the cool stuff. Most of us think that the coolest gadgets that a human can possess in today's world is nothing more than a damn 17-inch Apple Macbook Pro with Intel's hyperthreading i7 core processor that costs more than a damn car. So, I have decided to reveal an entire world of existing technologies to you; the real cool stuff that will make you wanna abandon your lowly life of blackberries, apples, windows and vegetables. You probably won't believe the fact that these types of technologies can actually exist (much less the fact they already do); but my friend, mistakes are a part of learning. So strap on your seatbelts and gear up to kick some very ancient and rusted arse.

Mooo(re)'s Law and the Digital Age

Let's see, firstly, how the world's most important invention came along and revolutionized science. The personal computer played an essential role in giving science its lover; someone who could sit and perform all tasks for us. Science could theorize, the computer could test; the perfect combination.

The first trials of a machine that could compute and calculate go to a 600-year ago Italy, by the particular hand of a man who was a genius. Every once in a bunch of centuries, a genius of a man like this comes waddling down from his hut and makes some of the craziest inventions and discoveries. This genius is normally far ahead of his primitive times and makes accurate predictions, theories and mechanisms that we end up using hundreds of years later.

One such genius was Leonardo Da Vinci; the same stud who painted the Mona Lisa. Da Vinci was a true "all-rounder"; a guy who got to paint, fly, design, build, think, watch, tell, deduce and do a lot more things to the very best of human ability (he was so talented that he outdid most of the geniuses in almost any field of science and time). In fact, he was so damn good, that historians think that no other human in the history of mankind can match this voracious intelligence. In the modern world, Da Vinci would be a combination of Albert Einstein, Isaac Newton, Stephen Hawking, Roald Dahl, Van Gogh and probably any other known talented person you can think of.

His hunger to become a know-it-all was so damn strong, that it made him do gross, serial-killer like things. Late in the night, Da Vinci used to creep out into graveyards and dig up corpses, which he used to slice open mercilessly and perform medieval post mortems, all so he could study the structure of the human body to make detailed sketches and notes. In short, he was the ideal 21st century student; something like what our teachers would love us to be.

Being a Christian, doing such a deed would mean that Da Vinci would get screwed by the Pope and sign his entry ticket to hell. So, to hide the blood on Da Vinci's arms and blunt the smell of rotting organs, Da Vinci's father used to give him a bath with red wine while the local Christian police came to each house searching for the traitor.

Anyways, what does a morbid, yet the most diversely talented, stud of a man have to do with the damn laptop I am typing on now? Well, quite frankly, Da Vinci has been rumored to be the first to propose the model of a working computer/programmable robot. Da Vinci was quite ahead of his 1452 AD age. So, while Italians back then were busy figuring out a way to spice up their spaghetti, Da Vinci was designing flying machines, computers and robots. No one ever seems to have found the designs for his proposed computer as such, but scientists have proudly demonstrated how Da Vinci's programmable robot would work beautifully if actually built today. Starting from Da Vinci, the concept of a computer sprouted around various parts of the world. Some of the smartest minds around the world wanted to create a calculator that could make hour-long complex calculations the work of a second.

Attempts to make a computer can be seen from the efforts of the Greeks and the Babylonians. In one of the ancient Greek wrecks, a device known as the Anthikythera was found. The Anthikythera was probably the coolest Greek invention because it predicted the positions of the planets and the moon extremely accurately; at a time when half the people thought that a thunderstorm was a result of the great god getting pissed off. In fact, Andrew Carol, an Apple software engineer, rebuilt the entire device using Lego (yes, that's how bored he was), and also predicted the solar eclipse of April 8, 2024 accurately, using this device. Probably the first model of an actual computer *computer* that actually worked was Charles Babbage's (no, it's not cabbage) Difference Engine; a truck of a gadget used to perform polynomial calculations and arithmetic. More progress on computer sciences was soon to come. A partnership between IBM and Harvard created the first programmable digital computer in the United States, the Harvard Mark I. Computers around the world soon went digital from analog (wires replaced gears) and before we knew it; Steve Jobs had come up with the first computer for the public.

Following all this progress, the massive wooden computers idealistically turned smaller, went electric and became more efficient. With the invention of the much needed transistor, the scientists finally dumped the massive vacuum tubes and bundles of wire that filled entire rooms. Computers adapted to a more portable sort of a room; a microchip. (A room as big as your palm! Funny!). So eventually, along came a portable computer, after decades of research and various massive rooms filled with wires and disks, we finally came up with something that could kick ass and tell the abacus goodbye.

Once computers went portable, they also started becoming cheap and available to all. More and more technology giants like Apple and Microsoft played heads on. They spent billions on researching stuff that could make computers cheaper, smaller and cooler. And following this process, one day, the computer became a household gadget from a rich man's prestige icon.

Sometime around then, Gordon Moore, who would go on to found Intel, came up with a standard law of computing. He said that computing power as of capability would double every eighteen months.

Which means, for example, if you have bought the best computer in town with the best specifications such as that of a fast processor and a good memory; your computer will become the next-to-best after eighteen months because another computer, with a better memory and a better processing speed will become commercialized. It is like saying that every eighteen months, computer scientists around the world will be able to fit even more stuff on an even smaller chip. So every eighteen months, Intel manages to pack more stuff than it has ever done before, onto the smallest chip it has ever yet made.

So, one day, following this senseless but very valid law, a tiny speck of plastic which is invisible to us might contain the entire world's data? That's more likely than physics becoming fun, for sure.

But this law, like all other boring scientific laws, has been created to make us innocent teenagers look like overgrown money suckers.

Why? Because, every time you spend a thousand dollars on getting the best laptop in town, it only takes 18 months for it to become the 'ancient' one. So, you now have to get another one. Well, if any adult or parent is reading this book, I would like to tell them that teenagers are not money suckers! We are simply following a law! A law has no damn exceptions! So the next time your teen comes and tells you he or she needs a new laptop, you can consider suing Intel or Gordon Moore instead of explaining the value of money to your teen.

But obviously, while having technology grow at the speed of light, we have a ton of advantages too. For example, the smartphone snuggled into your pocket has more computer power than all that NASA had when they chucked a couple of humans on to the moon in 1969. Imagine the millions of calculations that had to be done manually by the NASA scientists were all managed by a computer power that was similar or probably less than that of your damn Blackberry Curve.

So, at this rate of technological growth, say sometime around 2030, the current computing power of NASA will once again be confined to the power in our phones. NASA's computer power currently consists of 1,024 quad-core Intel Xeon processors and performance capabilities of 25 to 67 teraflops (trillion calculations per second), which is like saying that NASA uses the computers of over half the technology consumers in the USA. So, 19 years from now, the power of over half the computers in the United States will get rigged into our iPhones? Hopefully, yeah! Which means you will be able to watch over a hundred movies listen to over a thousand songs, talk on the phone with over two hundred people; all at one time. Pretty cool, eh? But this is where we hit the snag.

A lot of people around the world think that Moore's law is going to collapse. Which means that sometime in the future, it will take a longer time for Intel to pack in way more stuff onto way less space. Why? Let's see.

Firstly, our good old Moore friend came up with the law after understanding how Intel would make chips that would power computers. Moore said that with help from the perils of subjects, otherwise known as Electricity and Chemistry, we would make things on our chip even smaller than they were. A microchip is like a slab of cardboard, on which you have different parts that make a computer work. What Moore assumed, while making his law, was that every 18 months, a certain such part present in every computing device, known as a transistor, would become smaller and smaller and thus increase in quantity on the microchip.

A transistor basically makes sure that all the little parts in your cool gadget get just the right amount of energy (electricity), so those parts don't get burnt and stop working or so those parts don't completely run out of ammo. It's like saying that different servants in your house need to be fed different things to make sure they work. So, you a hire a separate guy who makes sure that these servants get their food demands. Without this special guy, your servants won't work all that smoothly because they wouldn't get their demand of food and so your cold coffee won't be all that nice. In simple words, this particular guy is very important to make sure your servants are smooth, well fed and therefore to make sure that your house is nice and perfect.

Basically, the more we have of these little things (transistors), the more precise amounts of electricity we can give to your little computing parts. This makes sure that they last longer while they also work much more efficiently. So, as our very useful little lads are becoming smaller and smaller, we are able to put more and more of these into our circuits which go into our computers. So, as we give the perfect amount of ammo to all the little parts in our iPhones, they work for a longer time. Not only that, but they also work much better when they get the precise amount of electricity. This means that they not only work much faster but they also work much better. So, as transistors become smaller, your gadget could fit in more of them. Now, because your gadget would have more transistors, it would be able to give all the parts in your gadget the exact amount of electricity they need. This would make your gadget work better.

Moving on, if you want a very fast gadget, it will have lots of other parts in it. If you have more parts in your gadget, you would also need more transistors to give these extra parts the right amount of ammo. So, as transistors become smaller, we keep adding more extra parts to your gadget in the same size, because the space previously used by the transistor is now being used by the extra part that makes your gadget fast. And that's exactly how Moore's law works. According to him, the smaller things get in our computing world, the faster the computer will get, because now we can fit in much more into that same small space.

BUT HOW DAMN SMALL CAN EVERYTHING BE?

We all know that everything is made up of tiny atoms. If you go beneath atoms, you find a whole lot more of boring physics. The point is, you can't build anything out of the parts inside an atom. The atom is the smallest scale to build stuff at. So what happens when your tinky transistors that we earlier talked about became so small that they were soon no bigger than a damn atom itself? What happens then? Well, for sure our phones would become a whole lot stronger but what happens after? What happens when we can't make our transistors smaller? Because transistors can't become smaller than atoms, our computing power remains the same for the rest of our lives! And it probably may even collapse, because hungry, idiotic consumers like us will want more and more of it. Imagine being able to play a million different games on your iPhone 20. You would obviously want to wait for an iPhone 21 on which you can play a billion games!

But that's when it hits you, there will be no iPhone 21 because the transistors can't be made smaller and the speed of your phone will remain the same forever. You will then, obviously get pissed and probably stop using computers and your phone out of frustration. What's the point? It's not like they're going to get any better; and after all, having a billion apps open at a time is much cooler than having a million apps open at a time. Then your friends copy you. A thousand other people copy your friends. Even more people copy your activity and BOOM. Soon, no one even wants an iPhone. The demand dies out, Apple starts making a loss. No demand, no supply and no Apples or oranges.

So, in very simple words, sometime in the near future, Moore's law may collapse (Yeah! One less law to learn!), and technology will still grow, but at a slower pace; until we come up with a new sort of material that can be made smaller, faster and better. Now that we know how computers may not grow all that fast in the future, we can say that the future of computing lies in other, different fields. But as we saw, the development of the computer and its present state was generated over hundreds of years. The ancient Greeks who came up with the Anthykyera didn't exactly think that their invention would be used to power a personal computer, a thousand years later. The computer wasn't something that came into someone's head one day and sat on his desk the day after. It was an ongoing and lengthy process that was needed to shape the personal computer.

So, now that we know how probably the world's most useful invention came about, let's glance at the crux of the matter; or the cool stuff that outdoes the very definition of coolness. The teleportation machine is somewhere midway in its development. If we compare it to the development of the computer, it has reached the "Harvard Mark I" stage. We know how teleportation would work and what remains to do is testing it on humans. Apart from the testing, we also have to find a way of making it cheaper. The invisibility cloak lies somewhere in that phase too. But as for the time travel mechanisms, we need a revolution. The day we can come up with something that travels close to the speed of light, we would have a semblance of a time machine. But keep one thing in mind. Great amounts of work in these fields have been done. We are far ahead in the field of teleportation, time travel and invisibility than any of us ever thought. So go ahead, and dive in.

Go Invisible (and steal some cookies)

You might find this hard to believe, but somewhere on this Earth, somewhere in Berkeley, California; in a lab filled with boring physics and pointless chemistry; folded and stacked neatly, lies a cloak that can turn its wearer invisible. Yes, the cloak already exists, it has been tried, tested and proven over a hundred times; and it works perfectly. The only problem is that it doesn't work perfectly on humans (well, duh). Scientists have managed to turn very very small objects invisible with this cloak, but that's where we hit the problem; the cloak doesn't work on objects bigger than the side of your hair. But a lot of bearded men around the world think that in the next two to three years, we will have successfully created a cloak that can turn humans invisible. In fact, some of the scientists even believe it might not take all that long for such a thing to happen, because if these scientists make this key breakthrough sooner, the invisibility cloak for humans will be invented sooner. Even though we can't turn us massive lads invisible, it is still pretty cool to see how an invisibility cloak might work. Let's take a look.

To get an understanding of how the invisibility cloak turns stuff invisible, let's see how we manage to see so much damn stuff anyway.

Tiny little packets coming from the sun surround everything on the Earth. These packets are called photons. When these packets hit objects and people, they bounce off and enter our eyes, allowing us to see these objects. The point here being if these packets were not there, there would be nothing that would bounce off other objects and thus enter our eyes. Therefore, we would not see anything at all. So, technically, if I managed to make sure that these packets bounce off me but don't enter anyone else's eyes, I would indeed be invisible to everyone else because nothing would tell these other eyes that I am in front of them. But how did I make sure that these tiny packets don't enter the eyes of all these other people? One way is to make sure that after these little packets bounce off me, I don't let them travel further. Once the particles hit me, I stop the particles from moving so they don't speed into other people's eyes. This will make sure that, for a start, no one standing directly in front of me can see me because the packets that are bouncing off me are not reaching their eyes. That's exactly how the cloak I talked about earlier works.

The cloak will make sure that these packets hitting you are stopped from moving or bouncing off. This cloak will act as Velcro, and it will make sure that these packets rushing towards you stick onto your cloak itself and don't go flying back. If the particles don't have the chance to bounce off you, they will obviously not enter the eyes of the people standing around you. So those people will not see you, but instead they will see all the objects behind you or around you, because the packets hitting these other objects are directly entering these people's eyes. And that is how the invisibility cloak will work. It will make sure that light literally bends around you.

Another amazing way to make yourself disappear is by using a hell of a lot of cameras and a hell of a lot of light. Researchers in China have figured out a very messed up way to get rid of your visibility. First, you gotta grab a piece of cloth, like a shirt or a coat. Then you take a couple of cameras and stick them at the back of your shirt. You then take a couple of TVs that you can fold like your shirt, and stick these TVs in the front of your shirt. Now you connect each one of the cameras stuck behind your shirt to the foldable TVs stuck to the front of your shirt; and power the whole system up. Then you switch on the cameras and the TV and BAM! The shirt looks like, very vaguely, it has vanished. Why? 'Cause there is no contrast between the background and the shirt. And this is so because the shirt is indeed showing the background itself (because the cameras are capturing the background and displaying it on the TV stuck ahead of the shirt).

To understand how that really messed up set up works, let's first look at how we see things. Say you take a white, blank slip of paper. You draw and cut your name out of this slip and then put it on another white sheet. When you try to look at this set up, you will find it very hard to read your name (oh really? I didn't bloody know that!). It's literally like your name has vanished into the white chart paper. But what happens if we take this same white cutout name of yours and put it on a black chart paper? Your name will glow like some turd has literally put lights on it and switched it on.

The point is that we are able to see a lot of things around us because of contrast. You are reading this book because the text on it is in black and the paper is white. This creates a contrast and allows you to read. But if both, the letters and the paper were white or if both were black, you wouldn't be able to make out the difference between the page and the text on the page. And therefore, you wouldn't be able to read. So, essentially, if we somehow manage to kill the contrast between your body and the clumsy background behind, you would become invisible. And that's exactly what our messed up Chinese invisibility cloak wants to do. By showing the background behind you directly in front of you, there will be no contrast between you and your background, and therefore to a normal guy, you would seem invisible.

Invisibility cloaks, bearded men hope, will be perfect for commercial use by 2030. So will I be able to waddle down to 9/11 and a grab a quick invisible shirt? No. 'Cause once these amazing things have been perfected, chances are very less that they will be given to the public. The army might want them and they may be illegally sold to celebrities and rich people with goggles on in murky nightclubs. But the chances are very less that the governments around the world will allow these amazing things to be sold to us common humans. Why? Well. How exactly would you feel if an invisible lad stole all your cookies? How would you feel if he barged into your house, stole your PS3 and went away? The point is, even if we manage to fully perfect an invisibility cloak, several men and women in robes will scream ethical insults to companies making these coats; and they will probably be made illegal, like drugs. An invisibility cloak will be more like a gun than a gadget. It will never be sold to the world, but rather kept protected and secret. But that doesn't change the fact that such coats will be super cool.

So what if we can't own an invisibility cloak? We can still teleport! And that, my friend, is a phenomenon that kills the coolness of the invisibility cloak. Let's see how.

Teleportation: When you're too lazy to use the metro.

"Beam me up, Scotty", yells Commander Spock (isn't that a brand that sells Lacrosse gear?), as his tiny assistant flicks on a couple of switches before the cheap visual effects of the early nineties come into play, swirl around a couple of humans before kicking them out of the massive spaceship and hurls them inside different planets.

We all love Star Trek, don't we? After all, they got the amazing fictional gadgets in town. They told us about long-eared aliens and about spaceships and stars. But that's not really true. Star Trek was in fact very very cheap. To save money on the set and instead of showing a massive aluminum foil ball of a ship land on the moon, they could simply make Spock yell "Beam me up" and show him teleport onto the Moon; which would save thousands of dollars on shooting costs.

Well nonetheless, this cheapskate little idiotic and deceptive way of saving money has led us to make a teleportation machine anyways. Sigh.

WHAT! WE HAVE MADE A DAMN TELEPORTATION MACHINE ALREADY?

Uh, yeah! And it's the coolest thing that mankind has ever made. Let's see how it works and what exactly it does (I thought a teleportation machine made me puke snails).

When we say that a machine can teleport objects, what we really mean is that it can make an object disappear over here and make it appear over there in a fraction of a second. Well, to obviously do something as crazy as that, we need to look at the atoms and molecules of this thing that we want to teleport. If we want to make the object move so fast, then logically, the only thing we can really do is to literally make that object move all that fast. So, how exactly does that work?

When things get really, really small, they also have a tendency to get really, really weird. Let's zoom down to an atom. Now, let's consider an atom to be a top that we can spin on both ends. The upper end of the top is called "up" and the lower end of the top is called "down" (innovative, eh?). Now, we can obviously, at one time, only spin the top on the "down" side or the "up" side, right? The top can't spin on both ends at once, can it? Obviously not. But with an atom, experiments have found that an atom always spins as up and as down, at the same time.

So, say that you have an atom and you have a way to measure the top's spinning side. This means that you have a machine to see which way the atom spins. The machine records the data from the atom and then tells it to you. Now, you take an atom and put it on this machine.

Several scientists, after having seen the results from the machine, have left wretched quantum science forever and instead gone to do sky-diving in the Bahamas. The results are so unbelievable that several physicists still remain in doubt about their pathetic choice of becoming a scientist. So, what is the damn result anyways?

The result tells you that half of the atom is spinning down and the other half is spinning up. Saying something like this in our world is no less than saying that Newton had blonde hair. But that does not mean that our results are wrong (although Newton was no blond obviously). What scientists still really haven't understood is how can atoms manage to do this. How can atoms spin up and spin down at the same damn time? Scientists have also found that atoms can practically exist in two possible states at once. If an atom was a person, they could be alive and dead at the same time. If an atom was a ball, they could spin right and left at the same time. But most importantly, if an atom was a teenager, they could be dumb and smart at the same time (pretty useful, eh?). Well that's weird. Hell, that's crazy weird. So scientists call these characteristics of atoms as quantum weirdness. And it is this weird little "tear-your-hair-out" theory that makes teleportation possible.

As I told you earlier, quantum entanglement is another really weird thing. If we have two atoms, and I make both of them rotate precisely towards the right, they will develop a sort of a relationship. Any changes on either one of these atoms will be immediately implemented to the other atom; no matter the distance. So, if I was to take Atom "A" and Atom "B", and then sort of tune them together (by making sure they rotate precisely in one direction), they would have a weird relationship. Then if I was to take Atom

A and put it on Mars, while taking Atom B and putting on the sun, they would *still maintain the goddamn* relationship. If I made any change to Atom A on Mars, say, for example, that I reversed the spin of this Atom, then Atom B on the Sun would automatically reverse its spin too. And this, my little friends, is the pathetic excuse of physics. No one knows how it happens, who does it or why such things happen; but they exist nonetheless. So along tagged a bunch of beasts, aka crazy evil scientists and created a damn machine that could use this principle to teleport stuff.

Well, this weird machine works in a weird way. To first understand how this happens, let's take three atoms; Atom A, Atom B and Atom C. First we take Atom A and Atom B and tune them together using our rotation principle (we spin the two atoms in a similar direction). Now, Atom A and Atom B are perfectly similar; like identical copies of each other. If I make any change onto either Atom A or Atom B, the other one will automatically and instantaneously get that type of change too. Now we take Atom C (which is a completely different atom, and is rotating in a direction different from that of Atoms A and B), and *entangle this Atom C with Atom B*. What would happen now? Atom B was entangled with Atom A, so it was exactly like Atom A in every way. Now that we have entangled Atom B with Atom C, Atom B has now become exactly like Atom C. But wait a sec, wasn't Atom B already entangled with Atom A? Yeah. So now, *Atom A is exactly like Atom C.* How? Because when Atom B, which was already entangled with Atom A, got entangled with Atom C (and then became Atom C's copy), it also made Atom A a copy of Atom C (remember, that because Atom B was previously entangled with Atom A, any change to one atom would immediately cause a change to the other atom). So, because Atom A has now got the physical properties of Atom C without having any type of direct relationship with Atom C itself, we can grin and say that Atom C has teleported to the location of Atom A.

So if Atom A was on Mars, Atom B on Venus and Atom A on Earth, we would have successfully teleported an atom from Earth to Mars in a fraction of a second, and saved ourselves a year's worth of travel and toothpaste food.

And this, my little bored teenagers, is the menace of teleportation. In fact, teleportation works so beautifully (in laboratories obviously), that IBM has also managed to teleport Calcium atoms to a distance greater than 5 meters. Every day, they increase the number of objects they teleport, rig in more energy and teleport this stuff further than ever before. Following this pattern, it's only likely that IBM comes up with a machine capable of Human Teleportation sometime by 2050.

And then my friends, is when the true fun starts. Once human DNA is successfully transported, that is without getting it mixed up with chemicals whizzing around in the air, the human teleportation would automatically fall in place.

There is however, a big "holy-shit-no-way" unknown factor to teleportation. What we are really doing when teleporting something is like sending the blueprints of the teleported object to another location, from which that object is built again from scratch. If we were to do something like this to humans, the human who wants to be teleported will have to die for a split second before showing up at the other location. And scientists are unsure if this will work so well with our body. But then again, all of these are mere thoughts, and what we do have today, is a machine that can teleport calcium and atoms. Tomorrow we will have a machine that will teleport humans, and that will be the most kickass invention of the millennium.

Time Travel: What the hell is time, anyways?

As I have mentioned repeatedly in this book, HUMANS ARE DUMB. The reasons for this include everything from our ability to make up crap and shove it into textbooks while the stupidity of what we do and say seems to play a major role too. Well, another reason that adds onto the madness is our inability to get a damn firm grasp on the nature of time. Time is money. Time is love. Time is pleasure. Time is life. But one question, still remains. What is time?

Newton thought that time was a constant little patty pet; which is to say he thought that clocks at different areas in the universe ran at the same speed. According to Newton, time could not be reversed, changed or even, as a matter of fact, studied appropriately. To Newton, time was the most amazing entity, something that was above everything, almost godly.

But then, along came another bearded genius by the name of Einstein and showed us that time is probably the weirdest thing we have ever come across. Einstein said that the entire universe contained an invisible substance known as "Space-time", and it was this substance that determined how fast or slow time moved. (Yes, this is the very same Space-time mentioned in Philosphysics).

Einstein theorized that this Space-time substance was like a liquid. This property made different objects in the universe have a different effect on this substance. Think of a massive swimming pool as our universe. Different people, fat or thin, enter the swimming people. While swimming, the fat people obviously create many more ripples than the thin people. Similarly, massive planets and bodies such as the Sun have a drastic effect on the substance of Space-time. They are the fat people of the Space-time swimming pool. The Sun creates such massive ripples in this swimming pool that it causes our entire solar system to exist. The ripples make space for the planets to orbit around the Sun in. In fact, gravity is nothing but different ripples in the swimming pool of Space-time. The Sun does not really create ripples in Space-time, (because if it did so then the Earth would be pushed away from the Sun), but really it just moves the Space-time around itself in different shapes and structures. How? It's the same concept as that of a ball. If you take a cricket ball and put it in the middle of a thin plastic sheet that is hanging above the ground, the ball would slowly create an inundation into the plastic sheet. That is, the ball would slowly sink into the plastic sheet and create a slope inside the sheet. And this slope, my friends, is caused due to gravity. If the thin plastic sheet is the Space-time and if the ball is the sun, the slope created by the ball is what makes the planet move around the Sun and keeps the solar system the way it is.

To understand better what I am trying to say, let's look at the swimming pool once again. Now imagine that the entire pool is called a universe. All the water in the swimming pool is called "Space-time". To show planets in the universe, we have tiny green cricket balls. To show massive stars in our petty little universe, we have massive footballs. Now, if I want to make a solar system in this universe of mine, I take a massive football, (the Sun) and a couple of tiny green cricket balls (planets). First I take the massive football and throw it into the pool. Then I take the small green cricket balls and sprinkle these around the massive football. If you observe carefully, you would see that the green balls are slowly moving towards the massive football. This happens because the big football literally makes the water angle down to a certain degree. This certain angling down leads to the other cricket balls coming nearer to the massive football; just as it would happen in any other situation. If we think of the water as a massive see-saw, this concept becomes very easy to grasp. If I take a massive football and several other small cricket balls and place them on a huge balanced see-saw, without letting the see-saw tip. I glue the football onto the see-saw to make sure it doesn't move. Now if I push the see-saw so it tips in the direction where the football is present, the entire collection of the green cricket balls would rush towards the football. We can also do this experiment on a very fluffy bed. If you take something very heavy, like a paper weight and put it on this fluffy mattress, the paper weight will literally sort of sink inside the mattress. This will cause the paper mattress to go down towards the paper weight and create a slope around the paper weight. Now if I was to do this same thing with at least ten cricket balls around the paper weight, the balls would automatically travel and get pulled towards the paper weight because of the slope that has been created around the heavy ball.

That is exactly how gravity works in our world too. The mattress in this case is the simmering pot of Space-time, and the various planets and stars around us are balls that literally sink into the Space-time mattress, causing slopes to be made around themselves. These slopes then pull down other planets and stars towards them, causing solar systems and galaxies to be formed.

So now that we know what Space-time is, and how gravity occurs because of changes in the Space-time fabric, it's pretty easy to figure out exactly how a time machine would work.

To completely understand time travel, we have to forget about our normal view of time. If anything at all, time is not something that only makes the clock tick. It can be minutely understood that way but that's it. When we look at the kickass-ness of a time travelling machine, we have to consider time to be a substance; like water. According to Einstein's laws, time is woven with space, so the Space-time fabric is actually what makes time happen. Now, if there are ripples or distortions in the Space-time fabric, time would "go" slower. That is, every clock in that particular area would tick slowly; your body would move slowly, the atoms down there would vibrate slowly. And you wouldn't feel a tiny ounce of a difference, because everything around you and in you is moving exactly at the pace of the slow-time. Something like this happens on Earth. As we discussed earlier, the Space-time material can be bent, distorted or changed by bodies such as the Earth and the Sun. So, quite obviously, the Space-time surrounding the Earth and the Sun is distorted or messed up. If the Space-time mattress is messed up, then time in that particular area would pass at a slower rate; so someone on a smoothly lined piece of Space-time would experience time at a faster pace when compared to a person experiencing time in a place where time is not very smooth.

So, say that I flew out to a place in the universe where the Space-time fabric is smooth, and I stayed there for about a month. When I get back to the Earth, two very weird things would have happened. Firstly, less than one month would have passed on the Earth from the moment I left. Secondly, I would actually be older than all of the people on the Earth, because for me and my body, more time would have passed as compared to an Earthling. To understand how these awesome things happen, let's forget about getting onto a rocket and zooming away to eternity and consider daily life examples.

Humans time-travel into the future on a daily basis. To understand how that happens, we are going to have to dive into a lot more of the stuff that Einstein said. And this stuff, my lazy little friend, can be completely mind-blowing.

Time Dilation

To understand Einstein's bizarre theory, let's consider two spaceships. Both these spaceships have synchronized light clocks inside them. A light clock consists of a cylinder which we can see through and a single ball that is bouncing inside the cylinder to detonate a second. The clock's second ticks only when the bouncing ball completes one round of bouncing successfully.

Now, if I was to keep one spaceship stationary and speed the other one, do you think there would be any difference in the reading of the light clocks? Well, If I had asked Mr. Newton the same question, I have would have got a yes, but according to Einstein, the answer is no. Einstein's theory says that time passes slower inside moving objects as opposed to the normal rate of time passing on still objects. Why? Let's take a look.

As we have learnt repeatedly in the perilous subject of mathematics, perpendicular distance is the shortest. That means that if I want to travel to the other side of Earth, I would reach faster if I managed to drill a hole through the center of the Earth and moved over to the other side, as compared to flying around half the world in an airplane. Following this same godforsaken law, the ball inside the light clock that we have on the two spaceships is bouncing in a perpendicular distance; which is the smallest possible distance it can bounce on. But the minute I start moving the clock, the ball would have to take a slanted course to maintain its bounciness (because of inertia, or resistance to movement)[1]. And because this slant course is not a perpendicular course, it is quite obvious that the course that the ball takes is not the smallest course. So, if the new course taken by the ball is not a slant one, it is a longer distance that the ball has to travel. And because the ball has to travel a longer distance now, it will take a longer time to do so, right? So the "tick" sound that you will hear from the moving clock will come at longer intervals, because the ball is now taking more time to bounce off the two ends of the cylinder. Therefore, can we successfully say that time is passing slower on the moving clock? According to Einstein, the answer is a yes. But the yes has certain conditions that we have to stick to.

Firstly, if I was in the moving spaceship itself, I would feel no difference in the pace of time. Everything to me would appear to be perfectly normal and nice. And why is that? Because as time slows down in the spaceship that I have taken, EVERYTHING in accord with the speed of time changes. My heart beats naturally slower to suit the slowed time, my brain thinks in slow motion and a lot of other things happen in slow motion.

In simple words, everything happening in the moving spaceship happens in slow motion; which eventually leads to the effect being equalized. This is the same thing that happens when you travel in a car and see a car travelling at the same speed right beside you; as if it were stationary. Because everything in both the cars is moving at the same speed, it appears nothing is moving at all or that the car beside you is still. Similarly, because everything in the spaceship has slowed down to a certain extent, it appears as if time is still jogging normally, when in fact it is doing exactly the opposite.

So, in short, the spaceship that is moving would have travelled into the future. Why? Because we would experience the notion of time at a slower phase while the spaceship itself would experience time faster. So, say, our two years of waiting and watching the spaceship, might be equal to only two minutes of the people inside the spaceship; because those guys have experienced time at a slower pace. The clock on that particular spaceship has ticked slowly, time in that spaceship has passed in a slower medium so the people inside that spaceship would be two years younger than anyone on Earth. And this phenomenon, my friends, is time travelling.

In fact, we do such time travelling all the time. Every time you get onto the metro or into your car and zip ahead of the hundred other people who are still, you are travelling into the future by extremely small time units. That is a snag to this entire time travelling business.

If our cars and planes were capable of moving at near light-speed velocities, then time travel would be a very common occurrence. Every time you got into your car and zipped past at the speed of light, you would have time travelled.

The trick to drastic time travelling involves reaching the speed of light in a sort of vehicle. This would ensure that you travel far ahead in the future and have the ability to become immortal.

Another great example of time travelling is looking at the stars. Some stars that are thousands of light years away from us are only seen by us hundreds of years later. Which means that when we see the night sky, the stars and the galaxies present in the sky are observed as they were a hundred or so years earlier? Yeah. This is so because light takes a hell of a lot of time to come from distant stars to the Earth. It takes over a hundred years to travel a distance of a million light-years; and when it reaches our eyes a hundred years later, we see how the star was a hundred years earlier. During the time when the light from that star travels to us the star might have changed, but we won't know or see the changed star until the next hundred years.

So, an ideal time machine would be a vehicle that can travel close to the speed of light. Then we could get into the vehicle and circle Earth at half the speed of light and travel several years into the future. But this is where the problem comes in. How far are we in terms of developing all these cool machines? When will we be able to teleport fly or sigh and not die? For all these inventions to actually kickass, we're going to have to compound our intelligence. That is, these technologies are so complicated and tough to figure out, that we're gonna have to fuse our brains with computers to understand how all this stuff will work. It's like saying that your iPod needs more memory to get those thousand more songs. But how can we, ehh, rig the brain up? I mean, there's no USB connection or anything up there to help us out. So, let's sneak a peek at the most complex piece of jelly in the entire universe - our brain, and look at how understanding this weird thing and uploading the whole damn thing to a computer will let us be a million times cooler.

Chapter Four
Smelly Jelly and Other Perils

The human body is tiresome, slow, complicated and gross. Don't get me wrong here, it works well, and better than the best of the machines we have built today. Especially that one very special, squishy and squirmy jelly-like organ that rests in the skull, otherwise known as the brain. In fact, it is thought to be the most complex assembly of chemicals in the universe. It is our brain that makes us conscious or aware of ourselves. It is that squiggly organ that separates you and makes you more intelligent than that chair you are sitting on. Without the brain, life as we know it would be impossible (for humans and octopuses at least). The brain is so damn complex, that even the most bearded geniuses have a tough time figuring out how the brain kills the fastest computers and most intelligent robots at thinking and computing and being intelligent.

But the greatest question that tickles scientists happens to be: How can a three-pound jelly organ question its own existence in the universe? How can a three pound jelly organ be aware of itself? How can a three pound jelly organ find math a boring subject? The point is, humans don't understand what makes the brain different. Other organs in our body like the liver and the stomach digest our food. Our kidneys help us use the toilet. Our heart beats blood around our body. Our brain makes sure that all these parts work fine and respond to what we refer to as stimulus. But all this while and during most of its pathetic lifetime, it also questions its own existence in the universe. While one organ is busy helping you take a crap, the brain is questioning its existence in our vast universe. Seems, a bit fancy and different, eh? What in Merlin's unshaven beard makes the brain different anyway?

The brain, in the days of old Greek-speaking men, was a subject of great debate. Some thought that the mind and brain were two different organs and things, while some thought the mind and brain was a single little slimy thing. Some thought that everything we do in our lives, all our actions and words, can be explained by studying the brain, while some thought a 'soul' controlled the human body. Well, very frankly, it would be true to say that we haven't gotten a long way since then. Sure, we have studied more of the brain than ever before, but we still can't precisely answer the question: "What gives us consciousness?"

The modern English-speaking bearded men think that it is only the damn brain that gives us "consciousness" and makes sure that we know we are alive and happy. They think that after having killed plenty of mice and hamsters, our brain is the only jelly present in our human body that gives us intelligence or tells us that we are alive in this tinky world of ours. Most people snicker at the fact that there is a "soul" snuggled somewhere deep inside. In fact, these people are so damn lame that they make jokes like "there is a damn soul under my foot, not in my body". The point is, no one can really get up, clear their throat and say that the brain is the only thing that gives us intelligence or consciousness because several other experiments have indicated otherwise (discussed in the last chapter).

The real reason that we have the most complex assembly of chemicals known in the universe stuffed inside our head is, as Mr. Daniel Wolpert, a revolutionary Cambridge neuroscientist says: because we need to move.

Think about it. From the second that we are born to our last breath, all we really do in life is move. To do anything worthwhile in life, we have to *move,* in some way or the other. Be it the wrong answer for math that you blabber out to a professor or as you try to gape at your partner's exam answers; you have to move in some way or the other to interact with your surroundings. Our lips move, allowing us to speak. Our arms move, allowing us to throw (our mathematics textbooks outside the window). Our legs move, allowing us to run (away from studies, with not a page of homework done). The point is, our only real way of communicating with the outside world is through moving (unless you and your damn girlfriend have that "cute" mental connection, that is). So, says Mr. Wolpert, we really need our brain only just to move. We don't need it to solve equations, learn formulas, understand theorems and dissect frogs. We only need our brain to help our body mobilize. Simple. Don't believe him? Look at an example yourself.

The Brain Eating Chunk of Plastic Pipes (That can also live forever)

Sea squirts are those weird little tube-like bottles that are glued onto sub-marine rocks. They look like small sections of plastic that have been cut out from a large pipe. These squirts, however, are pretty cool creatures. During the first phase of their life, they swim around in the sea, wink at female sea quirts and act like studs. However, when they enter their thirties and forties, they glue themselves onto a chunk of rock, and then they gobble down their own brain. That's right. They get rid of their brain by eating the damn thing and then giving out a nice little bubbly burp. To heck with good manners.

Why do they eat their brain? Because they know that from the depressing moment they glue themselves onto a rock, they will not move again for the next couple of hundred years; or their entire remaining life. A sea squirt can literally live forever, like a jelly fish, and it manages to do so by digesting its brain and not moving an inch instinctively for the better quarter of its life. "Why do we need the damn brain anymore?" they say, "We only had the damn jelly to help us move!"

The real point of this whole gross little encounter is that, a sea quirt that digests its brain at the age of 30, lives the next thousands of years without a damn brain. And it lives in perfectly good health. Amazingly weird? Humans, who have the most complex nervous system on the planet, are unable to live for more than seventy eight years, whereas an organism who has digested its own brain can linger on for a couple of thousand years. Wow!

This rather gross fact also supports the fact that the jelly stuffed inside our head is only to make us move. After all, it is this movement that helps us interact with the petty little world around us. Movement allows us to speak, write, text, read and maim.

But we're too damn big, complex and smelly to linger on without the brain. You see, the sea squirt and jellyfish are primitive creatures. Their entire body just consists of a massive, flimsy tube with two holes in it to pass water in and out. These tubes haven't evolved at all. As a sea squirt today looks and functions like a sea squirt that lived a couple of millions years ago. Also, squirts aren't intelligent enough to question their existence in the universe. Neither are they capable of sending their friendly chunks of pipes to the Moon. A human today has grown exponentially from a speck of chemical that existed several million years ago. A sea squirt has practically no organs, no hair and no fetish for food.

Humans, as opposed to those coolio chunks of pipe, have too much of an interest in everything. We eat more than we are meant to, sleep more than we are supposed to and don't give a damn about our rusting hands and legs. So, we need a brain to sort stuff out and help us live; or as Mr. Daniel would put it, we need a brain to move and interact with our surroundings. The brain helps us to be the smartest creatures on this rock; so why eat it?

Now that we know why the brain has been mercilessly stuffed into our cranium, let's see how exactly does it manage to keep you alive while dealing with the perils of math and science (mind you, which is a too damn tough job to do).

A brain, like all the other organs in our body, is a specific collection of cells. These cells of the brain are known as "neurons" (yes, it does make a good cuss, for example, "Holy neuron, what the crap just happened?"). A neuron, in its upper region, contains slimy, tentacle-like hair that stretch out. These are known as dendrites. In the middle of the neuron, a thin wire-like structure known as an axon stretches on and then branches into several hair-like projections which then go and connect with more dendrites of more neurons. Assemblies like these stretch all over your body and are active at almost all times, telling your organic machinery what to do. A neuron can have over a million dendrite connections like these. Two neurons, however, are never really sort of physically connected. Between an axon branch and a neuron, there is a tiny gap known as a synapse. It is through this tiny gap that a chemical travels and activates the next neuron in the line, causing a chain reaction to occur. Messages that whiz around our body in neurons work like these. A neuron can be thought of as a converter.

Say that we have three neurons in line, Neuron 'A', Neuron 'B' and Neuron 'C'. Neuron A receives a chemical that tickles its dendrites. The giggling dendrites then convert this chemical into electricity and pass it along their axons. When the electricity signal approaches the end of the neuron, it once again turns into a chemical to travel through the tiny gap between Neuron A and Neuron B. In this way, neurons convert electricity into chemicals and chemicals into electricity for the better half of their miserable life.

When it comes to the full working of the brain, we can still scratch our heads and go to sleep because we don't have a complete and thoroughly proven idea of its working just as yet. Two very strong ideas about the brain exist today. One idea says that each and every function of the brain has been localized, or has been given to a particular part of the brain to carry out. The other idea says that no specific function is performed by a specific part, but the entire brain in itself does all the work as a whole.

In years of research and medical progress, bearded geniuses have indeed found that certain areas of the brain co-relate to certain functions; but they have also found a major connection between all these parts. For example, no scientist can really say that only the 'x' part of the brain is responsible for storing memory and the 'y' part of the brain is responsible for seeing colors. Even though particular parts of the brain have been allotted jobs, those parts can't function and carry out their task individually. All the parts in the brain have an individual task to do, but also somehow manage to work together as a whole to give us our "conscious" experience. Scientists have yet to figure out this tantalizing relationship between these different sections of the brain and how the entire brain functions as a whole to be "conscious". Nonetheless, we do know some basic stuff about the brain's parts and components. Let's take a look.

Exploring a Mixture of Jelly

So our journey into the little brain starts right at the back of the head, at a region where our spinal cord connects into the brain, a region known as the medulla oblongata. This little region here makes sure that your heart is always beating at a constant pace; your blood pressure isn't shooting through the roof and basically makes sure that the essential stuff that keeps you alive is fine. If we somehow manage to stop the functioning of the entire brain of a person but keep the medulla oblongata of that person working fine; we will have turned that particular person into a petty little zombie. The zombie will be a living creature, its heart will beat and hands will move; but the zombie will feel no pain, emotions or remorse because the areas of the brain that give this feeling to the brain simply won't work. (Something of what we tend to become during exam days).

Moving on from our oblongata, we arrive onto the pons; a dude that makes sure we always walk and run in a coordinated way. Haven't you ever wondered why a zombie walks like someone has glued wooden floorboards on its calves and thighs? Well, it's cause his pons don't work the way they are supposed to, which in turn kills the coolness of coordinated movement in the zombie.

As we move further up in our skull, we come upon the cerebrum, a chunk of the brain resembling two massive walnut-shaped pieces. These walnut-shaped components contain some essential areas of the brain that help humans stay alive and active. The right cerebrum (or the right walnut) of this big nut controls the left side (left-side muscles) of the human body, and likewise for the right side. Each half of the entire walnut is also divided into several lobes based on the similarity of the functions performed by that particular lobe. The numerous little parts inside these walnuts and "lobes" do everything from giving you your laughs to making you purposely forget about the biology homework leading to eventual detentions. Areas in your brain that perform certain similar tasks are tagged as different lobes. For example, all the parts in the known human brain that help us to move in a coordinated manner will be grouped together as a lobe that will then collectively be given a name.

The various parts inside different lobes are cool. One of these parts inside the lobes includes the hypothalamus; which helps keep a tab on your memory. This particular part checks your hormones and also keeps your energy supply constant. Another part, known as the thalamus, stores most of your sensory information and also helps you think about that good ol' cheeseburst you hogged last Saturday.

Now, let's fire away the questions.

Firstly, if the brain is so powerful a damn machine, then why can't we multiply big numbers, predict various hands in poker or even memorize entire novels? If a computer can do all these things, then why in the world is it considered an inferior to the human brain?

To answer that, let's take a look at the "real" definition of intelligence.

Deep Blue's Stupidity and Kasparov's Intelligence

Sometime during early 1996, IBM came up with a massive computer called "Deep Blue". Deep Blue had been built to play chess and contained specialized computer equipment that helped it to do so. Bearded men at IBM claimed that Deep Blue could evaluate over 2 million chess positions at a time. So, to test the monster of the machine they had built, the proud old studs at IBM gave a call to the former world chess champion, Gary Kasparov and challenged him to a tournament with Deep Blue.

During the first round of playing, Kasparov won the tournament with a sly grin on his face against Deep Blue, even though Deep Blue managed to win two matches against Kasparov in the best of five series. Following the second challenge in 1997, Kasparov lost to an upgraded version of Deep Blue and got pissed at IBM because he claimed that a human operator meddled with the gameplays during the chess match, and thus said that IBM had cheated. IBM however, said, "You lost kid, forget about it". To piss off Kasparov even more, they denied his subsequent challenges to the machine and even disassembled Deep Blue.

Today, claiming Deep Blue was better at chess than Kasparov is extremely valid and true; because the machine beat Kasparov at chess.

However, if we were to compare the intelligence of the two; Kasparov would rank much, much higher. And that is not because Kasparov was a natural prodigy or a genius or a goddamn savant; but simply because Kasparov had a human brain instead of metal microchips. The brain beats any supercomputer in the world because it programs itself to suit different situations; as opposed to a computer being programmed by different humans to suit different situations.

For example, after the chess match in 1997, Kasparov got into his car and drove from the match venue to his hotel. In the hotel, he got inside the elevator pressed the button of the fifth floor. He went to his room on that floor, opened his door using his key and shut the door. Now, all this stuff done by Kasparov is not something we would call very "intelligent".

But if Deep Blue was put inside a mechanical body and was told to imitate Kasparov's activities after the match exactly as Kasparov had, Deep Blue would fail miserably. Firstly, the men at IBM would have to spend years re-programming Deep Blue so it could drive a car from the game venue to Kasparov's hotel. Then they would have to test this a hundred times before programming Deep Blue to identify elevator buttons in Kasparov's hotel. Then they would further program Deep Blue to make it move from the elevator to Kasparov's room and open the door. In short, the activity that Kasparov had planned and executed in 10 minutes would take over five years for Deep Blue to even imitate. After probably a thousand lines of code and many million dollars, would Deep Blue come close to performing the basic human activity that Kasparov had done so easily. Nonetheless, during all this while, Deep Blue could still beat Kasparov at a game of chess.

The true definition of intelligence, therefore, is not just about being able to beat a chess master. Rather, intelligence is the refined ability to react differently to different situations and stimulus; which

Kasparov, or for that matter, any human can do so efficiently. And why is this so? Because the human brain constantly re-programs itself, during each and every second of its life, to react appropriately to different stimulus as it experiences them. Hence, the ultimate computer machine would not be one that could calculate up to a billion numbers, but rather one that could re-program itself a billion times to respond to each and every stimulus that the universe has to offer. And after having the ability to do all that, if the machine still possessed the skill to beat Kasparov at chess, then we could definitely call the machine more intelligent than Kasparov.

The Lameness of the Jelly

The brain may seem all goody-goody at first glance, with a proud, smug face and the power of the world; but there are a couple of snags that compete with the advantages. The sheer quantity of these disorders, not to mention the number of people around the world who are diagnosed with such symptoms, drive experts crazy. For example, in a world of seven billion people, more than 40 million Americans have some sort of a mental disorder. Around the world, it has been estimated that around 500 million people have been diagnosed with some sort of a mental disorder. And hey, these 500 million are only the diagnosed ones who could pay for a hospital. There can be another hundred million living in places like Djibouti who have disorders too but can't afford a diagnosis. The point is, if so many people in and around the world have a goddamn disorder in the brain, then something is seriously messed up with our jelly, right? If so, then what's going on? Apart from the disorders, let's take a look at some other snags of the brain:

1. Firstly, we know literally nothing about the inner workings of the brain. Sure as heck, neurology has become a lot larger since the dawn of the cookie called Earth, but it still has a long, long way to go. Even though we know a great deal about what the brain is made of and how different parts of the brain work, we still lack knowledge of the specifics. For example, we still remain clueless about how the different parts in the brain

quickly merge the information they process to give us our conscious experience, as discussed earlier. If we don't know the confirmed details that the brain dude uses to give us the "oh-I-am-alive!" feeling, well, then we know nothing. Why? Because that fact stops us from getting rid of some of the most disgusting and cruel things that have ever been noted to haunt us evolved monkeys (as discussed in point two). Furthermore, it prevents us from understanding the most complex machinery present in the universe. Pretty blatant and sad, eh? It's also pretty messed up, because saying that we can't understand our own brain is no less than saying however smart our brain may be, it can't understand its own damn self.

2. Coming back to the issue of the brain malfunctioning and short-circuiting, let's speak of some disorders. Some of the most horrifying diseases to have haunted us lowly humans are related majorly to the brain dude. Think about it. Because the brain is the control powerhouse of our body, it has the explicit control over almost all of us, each and every damn centimeter. So, if this powerhouse was damaged or hurt, there would be endless possibilities of what the body could undergo because of the damage. Well, also some of the most messed up and "rolling-on-the-floor-laughing" disease/disorders in the world are caused by minor power cuts in the brain, and these are the ones that make the brain look like dust. Let's see some of these:

a. Cortard Delusion: However strong and kickass our brain may pretend to be, there are often times and cases when it behaves like the most pathetic and retarded thing to have ever come on the surface of the Earth. One such example of the brain acting like a dumb piece of rock happens to be while it's suffering through a disorder known as the Cortard Delusion. In the Cortard Delusion, a person's brain is damaged to such an extent, that he thinks that he is dead and that whatever he is doing after being dead is being done in a mysterious form of afterlife while its body is rotting away in a putrid coffin. We still don't have a very clear idea regarding how this works, but when someone's going through this weird disorder, they assume that they are dead and that their body is mysteriously decaying in a coffin whereas their godforsaken soul is wandering around in the real world, having the time of its *after*life. Not only is this insanely funny, it is also very very sad. Most people suffering with this syndrome have to go through painful lapses and seizures until their brain gives in and kills itself in a ruthless process. In short, this disorder (apart from being hilarious to watch) is idiotic and wrong. Its existence is wrong because it kills its victims slowly and painfully, while also making them believe they are already dead. Have you ever been dead and wondered why you are dying again? Trust me, the feeling is not so good.

b. The Fregoli Syndrome: This is one of those utterly sick mental syndromes that leave scientists scratching their heads every time they diagnose a patient with it. When someone is diagnosed with the Fregoli Syndrome, they tend to see a massive crowd of similar looking people around them. That is, to say, they can't distinguish between different people, because all these different people seem to appear as multiple copies of this one person they have seen at some point in their miserable life.

Say, for example, that one morning you get up diagnosed with the Fregoli Syndrome. The first person you see in the morning is your mother, serving you a glass of piping hot milk. Then, your mom leaves for her gym schedule and you decide to get ready for school. When you come into the corridor, ready for school, you see your mom cleaning the carpet. "Mom, didn't you go to the gym?" you ask. "It's me, the maidservant, not your mom, you duck", replies the woman who seems exactly like your mom. Throughout the day, you see your mom everywhere. You see her in the school bus, playing the PSP near the window and at the classroom, as the janitor. Sometimes, you even see two copies of your mom pointing towards you and snickering about how you don't brush your teeth. Seeing all of this screwed up stuff will obviously make you want to tear your hair out and then use it as a damn wig before the day even starts (no, that's not a typical symptom you would expect to pop up with the Fregoli Syndrome nonetheless, teenagers are known to do crazier things), but this is an existing mental disorder that literally drives scientists up the wall. Researchers are sure that The Fregoli Syndrome stops the visual-identity areas in the brain from working properly, and therefore cause them to essentially see the same damn person everywhere. But that's about as close we would ever get to understanding The Fregoli Syndrome (until we manage to crack open the stubborn brain).

c. *Somatoparophrenia:* Another one of these (literally) mind-boggling disorders happens to be what sounds like an ancient Greek curse word, Somatoparophrenia. As the name quite evidently doesn't suggest, Somatoparophrenia is a disorder that causes patients to believe their own body parts are not their very own. That means, one day you wake up, enter the washroom and start brushing your teeth but soon realize that someone else is brushing them for you with their hand and not yours; and then a light bulb flickers on somewhere and you figure out it's a random arm attached to your body that is brushing your teeth for you. You are utterly sure that it is not your arm because you can't even sense the damn thing while it clumsily swipes Pepsodent on your yellow teeth. How? Because your brain thinks your left arm doesn't even exist, and therefore it disclaims the very presence of that arm. While a random arm brushes your teeth to eternity, your brain just assumes that your Mom is standing next to you, scrubbing your white diamonds.

To understand why the brain acts like such a dumb donkey, we have to talk about the concept of brain maps. Brain maps are like a pictorial representation of your body wired into the system. The brain uses these maps very often to move your body without having to look at the different parts of your bod. For example, when you want to move your arm to the left, you simply move your arm to the left. You don't have to look at your arm with your eyes while moving the arm, and that's because your brain already knows where your left arm is.

These brain maps make sure that your brain has a fairly good idea about where the different parts in your body are at present. So, when you get affected by Somatoparophrenia, a certain body part of yours gets erased from these brain maps. For example, say, that your brain accidentally erases the position of your left arm from the brain map of your body. Your arm will still work as well as ever, but your brain just won't know that your left arm even exists anymore because the damn thing has been marked off the record in the brain. So, the next time you use your left hand, your brain will think someone else is using their left hand to support you, or that you have glued on someone else's arm inside your socket. This can be a pretty grueling and messed up disorder because several cases have been noted where people have chopped off their hands and legs just because they thought they were someone else's. In summary, even though this entire disease may seem funny, it really is very disturbing and gross, just like the other ones.

d. The Alice in Wonderland Syndrome (Todd's Syndrome): If you wake up one morning, and find your coffee mug towering a couple of feet over you, with the teaspoon spanning twice over your dwarfed legs and your alarm clock competing with the Big Ben, be warned, for you have scuttled into Wonderland (if you sleepwalk, that is and presuming also that your house is often subjected to magical cyclones). Either that, or you are suffering from another one of those highly messed up brain disorders known as Todd's Syndrome, otherwise known as The Alice in Wonderland Syndrome (ingenious, eh?). As the name quite evidently *does* suggest; anyone suffering from Todd's Syndrome will lose his sense of scale to his surroundings, and go berserk trying to figure out what the hell is going on anyways. To blame for this wretched, stinking little disorder, we once again have to thank the brain, particularly the "lobes" we talked about earlier. The lobe that specifically gives us our vision is known as the parietal lobe, and when things start to get messed up in here, the brain loses its spatial capability and starts acting weird. In other words, it loses its ability to size things with regard to the body, and therefore thinks that everything is either amazingly big or very small (well, now we know Gulliver was not entirely mentally stable). Even though this disorder may seem like utter rubbish; it is a startlingly common happening in the world around us. Not only that, but this entire disease is extremely painful and torturous. The minute things start becoming big and small all at the same time, the patient also experiences a massive pain in his head; as if someone's set off a firecracker in his skull. Even though this particular disorder doesn't directly kill or maim, it's pretty idiotic, right? I mean the same brain that can do a million complex calculations to send a bunch of evolving monkeys to the moon, 38,000 kilometers away from our homely rock, also occasionally sees the Empire State Building to be the size of a hairpin? What the hell?

e. *Sleep Paralysis:* This one just does it. This disorder is like a signboard stuck on the back of the brain that reads, "I AM A DUMB COCKTAIL OF JELLY AND WATER". Why? Well, because this damn disorder gives you the worst ten minutes in your personalized hell. Like, literally, hell *hell.* Let's see what happens when someone is going through sleep paralysis.

Imagine that you wake up one fine morning, ready to jump out of bed and stretch your wrinkled body. But the minute you try to do so, nothing happens. Your body refuses to move or nudge at the wish of your impeccable brain, which mind you, is still lost in its petty dream. Then suddenly, you hear a hissing noise, and find a gigantic cobra slithering towards you, its beady eyes longing for your flesh. You try to move your paralyzed body but it won't shift an inch. You try to shout, but your mouth refuses to open. You wonder if you are still dreaming, but then realize that your eyes are actually open, and that the cobra is actually there. You can feel the quilt on your legs. This definitely can't be a damn dream. Then comes the bad part. The cobra prepares for the attack, arming itself with venomous poison and glaring at you, waiting to start its assault. Just as it brings its fangs close to your skin, you roll over on the damn bed and fall right on to the damn cobra itself; waiting to feel the stinging pain of the venom and then instant death. But the cobra is no longer there. It never was there. You look up, rubbing your eyes, wondering if you had dreamt it all, but you realize you are still on the floor, exactly where the cobra had been. What the hell just happened? Weird? A scary fairy hairy tale? No. *These are the symptoms of sleep paralysis.* If you are a victim to this disorder, then several horrifying things happen at once. First in our list horrors, you become paralyzed, and you can't move even an inch instinctively. Secondly, because a certain part of your brain is still dreaming, the elements from your nightmare can be seen by your awake and open eyes. So, even though your brain convinces itself that it's dreaming, your eye says, "What the hell, brain brother, you must be asleep but I am very much awake and open. I can see the damn snake right there". It's not like the eyes are seeing the snake actually, but they do send that sort of a message to the brain. All of this confuses the brain (it has rather confused itself). So while the brain tries to figure out what's going; if it's awake or asleep; if it's dreaming or not, the insiders of your nightmares approach you while you remain paralyzed and stuck to your body with your eyes wide open, waiting for the torture. Ultimately, the brain either decides if it's sleeping or awake or if it is dreaming or not dreaming and everything then tunes back to normal. But why does something as bizarre as this happen in the first place? The answer is quite simple, really.

To understand how sleep paralysis works, let's understand how sleep works. The brain, when you lie down to sleep, turns off your movement muscles in the body. That means the first thing that happens when you sleep, is you become paralyzed; you can't move. Well, most us need to get paralyzed while sleeping because if we tend to dream about running or moving or performing any activity, we would move our muscles and actually do the activity in our sleep, while we lie on the bed (sleep-jogging seems exciting, eh?). Secondly, the muscles of our body need to get relaxed. They are pumped up and constantly working through the day; so at night, when they become paralyzed, they carry out their repairs and rest. Ironically, while our body remains paralyzed, weighing onto the bed, our eyeballs move into a frenzy of retarded motion that gives us our dreams and nightmares. This is known as the Rapid Eye Movement (REM) stage in sleep, and happens only when we are dreaming. Then there is as second stage in sleep, when our body becomes mobile again and we stop dreaming. This is known as the No Random Eye Movement stage (NREM) (creative?). Normally, during the night, when we wake up and scratch our heads or visit the toilet, we are in the NREM stage. Our body constantly switches between these two states during the night. However, if we wake up when our body is in the REM stage (which rarely happens, because we are too deep in our sleep at this point of time), our body would still be paralyzed, and our brain would be smitten in a dream. And that's when things get messy, because it's like saying that half our brain is awake and the other half is dreaming. Well, sure as hell, I wouldn't want to wake up in my nightmare, so this is just another one of those massive glitches that come packed in with the jelly box. Such sort of paralysis has been known to have had a devastating effect on patients who remained traumatized once the nightmarish experiences were over. Some patients seem to have had a nocturnal death in their sleep paralysis because of the sheer shock they experienced. In the end, this sort of a thing is not all that pleasant after all.

Having looked at these highly retarded brain disorders, let's see even more things that make the pretty brain as inefficient as a mathematics textbook. We know that the brain contains huge amounts of malfunctions and glitches by the sheer amount of disorders and syndromes it has to offer, not to mention the craziness of these disorders, let's take a look at the computing limit of the brain and why the brain is such a smelly jelly:

3. Talking more about biological limitations; the brain is excruciatingly slow. The neurons that we talked about earlier transmit signals at slow speeds. These slow speeds are one of the main reasons why the brain loses the battle to a computer when calculating great numbers. In fact, the brain's circuitry transmits signals at speeds that are million times slower than the speed of the circuits put in a computer. Not only that, but it takes a long time for different messages to get to the different parts of the brain. It takes even more time for the neurons in the brain to come up with an answer; all of which makes humans seem sluggish when calculating. If the neurons transmitted signals as fast as a wire carries electricity in a typical Intel circuit, humans would be able to calculate these big numbers pretty quickly. Not only because the neurons leading to the brain will send signals faster, but also because the neurons inside your brain or the neurons your brain is made of, will process the information faster. This is also why reflex actions in our body are not sent to the brain. If you touch something very, very hot, the signal that tells your body that you have touched something piping hot only travels as far as your spinal cord, or your back. The spinal cord then sends a signal back telling your hand to move away from the hot object before you burn your skin. If this same signal travelled all the way to your head and back, your hand would be burned by the time you moved your palm and fingers.

4. Another limitation of the brain is the sheer amount of energy it consumes. Despite the fact that the brain only adds on 2% weight to the human body, it consumes more than 20% of the body's energy; which is a heck of a lot. The brain is a hungry beast; and almost all of the energy that we give to the brain gets consumed only in the firing away of the neurons or in the communication processes. So, even if we somehow manage to increase the speed that the electricity in the brain travels with, we will need a whole lot more energy to support something that powerful and fast. And that, my friends, is not as simple as saying that we will need to eat a lot more food. After eating a lot more food, we would have to change the way our body makes energy from food. So, if we want to make the entire brain

superfast through increasing the speed of the neurons, we are going to have to change all the goddamn blueprints in our body (which is quite a tiresome process, unless you're a woodie from the future, that is).

There you go. That's all the coolness that was ever stuffed into the brain, ruthlessly murdered. The brain may be a kickass organ but it has so many problems that we can hardly harness its true potential. It's like saying that you have the best computer in town, but you can't use it completely because there are way too many power cuts in your area.

Let's summarize. The brain may be smart, but its coolness diminishes because:

1. It, as of now, can't understand its own self.

2. It is way too slow to compete with a computer and perform huge calculations.

3. It malfunctions in bizarre, pathetic ways that affects millions of people worldwide.

4. It consumes way too much power for what it does - a hell of a lot of thinking.

Having seen the madness of the disadvantages that the brain has to offer, we are left with a challenge; to evolve. If we want to harness the full potential of the brain and leave out all the uncool stuff about it, evolution is our way out. We've been evolving ever ancient squirmy jellyfish appeared millions of years ago in the oceans, and hey, we became grown monkeys from a soup of chemicals.

But wait a damn moment. Natural evolution is slow. We have to wait for a couple of thousand years to see even a tiny hint of evolution; much less a completely efficient brain without any "tear-my-hair-out" disadvantages that it has to offer. So we can't really expect a perfect brain molded through evolution by the end of this century or so. We would probably see such sort of a thing sometime in the next million years.

Fortunately, mankind is now on a frontier where we can trigger our own evolution. We can make ourselves evolve faster with the "holy-shit" technology that we have and will develop in the coming centuries. And because we possess the ability to kickstart our own evolution in our own bodies through our fancy merchandise, we can overcome any time constraints that evolution silently sets up. Let's take a look at what we have in our stores to majorly upgrade the brain using our very own toy, technology. Let's see how technology will change in the future and help us make our brain better than it has ever been before.

Changing Technology and Making Jelly Computers

If we rowdy humans are anyway near to computerizing our brain or becoming mechanical cyborgs, it is because we understand at least some basic aspects and components of our three pound jelly. As Ray Kurzweil discusses in his book, The Singularity Is Near, our understanding of the brain gives us greater flexibility to do what the hell ever we want with it. So, before we even start considering the technology we will use to give the brain a digital dose of power, we need to talk about understanding the entire brain. Every bare inch of that jelly needs to be completely understood by bearded men before we give it any type of a shot in the arm. And as I said earlier, my little hairy friends, understanding the brain is no easy task.

One way to understand the brain and completely deduce it logically is by reverse-engineering it. That is to say that the brain can be disassembled and each and every area and lobe can then be studied in detail. One could also input this information into a computer model where one can run real-time simulations of human intelligence. To run these simulations of the brain, we first need to know enough about the brain, at least about all its parts, if not the exact software it works on. Till today, the instruments that we have used to scan the brain have been crude and inaccurate to say the least, and there were few organized methods to go about doing brainy stuff. Today, with better scanning capabilities and electronic machinery, scientists have been able to explore the working of the brain in real-time; that is to observe the brain when it is actually thinking or solving a math problem. Once we get this sort of raw data in hand, for example, which areas of the brain are activated when it solves a math problem, then we can process the complex stuff. We can make a computer model of the brain, feed into it all the data about the brain that we have collected, and then ultimately run a simulation of the human brain. This will give us a better idea of how the brain works in a daily-life situation.

Some of these cool techniques to explore our mass of jelly are being studied by some of the best bearded men around. There are several of these so-called techniques that can help us understand our brain, involving everything including slicing open the whole damn brain to poking the skull with tiny electrodes.

The Brain Dilemma

Dr. Ramachandran is one of the leading neuroscientists around the globe today. He's one of the coolest scientists to have ever walked on this balmy surface of the Earth because unlike other scientists and neurologists, VS Ramachandran likes to keep things very, very simple. As VS explains in his groundbreaking novel, "The Phantoms in the Brain", we can gain a great deal of knowledge about the functioning of the brain by studying the disorders that the brain ever so often comes up with. Ramachandran describes his dealings with patients diagnosed with the most surreal brain syndromes, and how, upon offering a treatment to these syndromes, there has been major progress in the gaining of neurological knowledge. Ramachandran's techniques are often easy as hell, but they solve the most retarded and complex brain problems that have been troubling our petty brains for quite a while.

Sometimes, brain damage or injury can actually be an amazing feat.

Some people have been known to go around banging their heads and getting amazing powers instead of just dropping dead.

The Brain Dilemma - Tommy McHugh

A (literally) mind-blowing example of someone like this is that of Tommy McHugh, a 35-year old guy. One fine day, Tommy, a man with a history of drugs and alcoholism starts hearing a weird sort of ringing in his ears before his head explodes with a searing pain from the inside while he is using the washroom. He immediately goes to the doctor, who frowns and tells him that his brain has started bleeding from both lobes because the blood pipes in these lobes have ripped apart. So, Tommy goes into surgery, gets his jelly back in place and strapped on before getting back to his usual life. And there ain't no problem yet. But for the first three months after surgery, Tommy literally goes crazy. He speaks in a weird rhyme all the time, which was anything but fine. He starts behaving awkwardly and weirdly, acting as though had someone turned off a switch in his primordial brain. So what happens after?

Well, sure as hell, Tommy would have made a great rap star with godforsaken rhyming powers, but his now damaged brain had different plans for him. After having had brain surgery rip his brain apart, Tommy McHugh becomes a goddamn artist. Yes, that's right, the guy whose brain almost bled the crap out, then suddenly gained a weird way to paint like Van Gogh. So Tommy picks up a brush and starts painting in rooms, on books, sheets and every damn thing he can lay his hands on. He paints aliens, weird avatar-like blue men and weird horses that look like they've taken a bit of a mathematical dose. His drawings are amazing, and even sell for thousands of dollars, making this young little Tommy wannabe-Hilfiger man a success. Good story, eh?

So what exactly did happen? What made Tommy's brain go absolutely nuts and turn him in to a Da Vinci from a drug addict? Well, when Tommy's two blood vessels burst in his brain, doctors zipped up everything with metal coils and clips to stop the bleeding and to keep things going in his brain. Then they sewed his head back on and told him to scurry off. Well, what happened in there after they sewed in chunks of metal in his brain, we don't know. What we do know is that if we scamper around the brain in appropriate ways, we may discover even crazier things; stuff that is completely bizarre. There are a couple more cases like this one in the world of bearded geniuses, and even though they don't really help kick some neuroscience ass, they do help us figure out the localized areas of the brain.

Optogenetics and Controlling the Brain Using a Torch

Another modern and 'wow!' method to observe the brain's activity is through Optogenetics. Optogenetics is a relatively new field that came into play in the recent past. It is such an important field because not only can it record and show the activity of the brain at a particular time, but it can also be used to control the brain to an extent. So, how exactly does this kickass method work?

Well, if we consider neurons to be wires in our brain, then the brain happens to be nothing but a massive tangle of wires. Different parts in this tangle of wire control different stuff, but all of these wires contain electricity. The electricity in these tangles is controlled using a different variety of switches (the same switches that turn your fans and your lights on and off). So, different switches in this tangle control where the electricity in the entire tangle flows to. What Optogenetics does to control the brain is; it tells it which switch to activate when and hence turn on, allowing electricity to flow through a particular wire in this tangle.

In biological terms, Optogenetics works by using a special type of protein that tells our little buddies, neurons, to fire their signals away. This protein is characterized under a broad term known as an Opsin gene. There are two very special factors about the Opsin gene. Firstly, the Opsin gene can instantaneously activate the neuron it resides in, that is to say, it can fire on the electricity in these neurons. Secondly, these Opsin genes are light sensitive. Which means that these genes only cause the neurons to fire on when light is shined onto them. These two factors make scientists go crazy about Opsin genes. In fact, some scientists even proposed getting married to Opsin genes but gave up on the idea after thinking that they would have to spend the rest of their white-coated, bespectacled lives with a couple of test tubes.

But why are scientists so happy about using Optogenetics to control the brain? I mean, other methods like electrode shocking and drugs have existed since forever, and these offered brilliant ways to activate different areas of the brain, so why go into a frenzy over Optogenetics?

Optogenetics are different, as the very founder of Opsin genes, James Crick says, because they offer ultimate the precision in activating the brain, and do so immediately. Unlike electrodes, that activate entire areas of the brain when switched on, Optogenetics only activate certain, controlled areas of the brain. Also, unlike the existing drug methods to activate the brain, Optogenetics ensure that the entire process of activating sectors of the brain is made increasingly fast and accurate. And clearly, would you prefer getting hit by a small beam of light or being painfully electrocuted by massive, metal electrodes that look like they have been whisked out of Frankenstein's laboratory? The latter sounds better to me, at least.

So, how exactly does your ideal Optogenetic experiment work?

You take a nice fat hamster named Fred. You then inject Fred with the two Opsin genes that travel to his brain and place themselves in areas of his brain. One of these Opsin genes goes and snuggles into the area of the brain that helps Fred recognize and distinguish smell. The other Opsin gene travels to the area of the brain that helps Fred see and recall visual stimulus. Then you take this fat little Fred and put him in front of a massive maze with several entrances. One of these various entrances into the maze has cheese in it and the other entrance has a signboard saying "Exit through here".

Now, if we assume that our hamster can read a signboard and has a strong sense of smell, we can definitely say that the damn hamster is also confused. Because Fred can smell the cheese and read the signboard as well, it (I'm sorry, I mean "he") can't choose between the option of eating the cheese or getting out of the maze, so it starts sniffing and waddles around both entrances, trying to make the decision of his life. Then, we cunningly grab our torches and decide to shine light onto the hamster's brain. First, we shine light onto the Opsin gene that is in the smell-detection area of the brain. The Opsin gene causes a neuron to fire away in this area, which shuts down this part of the brain or temporarily switches it off. Now, the hamster can't smell the cheese because the brain's sensor that detects smelly equipment is conked out. Therefore, Fred just looks at the signboard and goes through that entrance with a grin on his face hoping to find the exit. If we decide to shine light on the Opsin gene put into the visual area of the brain, a neuron switches off this visual area of the brain. Therefore, Fred would take the cheese path and get lost in the maze because temporarily it can't process the information related to its sight. In this way, evil humans test the cognitive abilities of various humble creatures and figure out loads more stuff about the brain in the process. Obviously, what actual bearded men do is way more complex than playing around with a hungry hamster named Fred.

By using loads more of Opsin genes and way more stray hamsters, scientists can piece together the function of each and every part of the brain.

So, what exactly happens once bearded geniuses get together all their data? Assuming that we use Optogenetics and other cool techniques to explore the brain, and manage to download all its information and software on our desktops, what is it that we do next? What happens when we have modeled each and every neuron and understood it completely? We then rig all the data in to a computer system and turn on the switch. We then see a computer act like a human, make human choices, make choices that are conscious and not defined just by algorithms. In short, we have uploaded an entire brain into a computer; piece by piece; neuron by neuron. But wait a second. How much hardware do I need?

Getting All the Data Together: Software and Hardware

As Ray Kurzweil suggests, the brain can perform around one quadrillion calculations in a second.

(We must note that, at this point in time, the brain computes faster than the computer because it can process data parallelly or perform calculation simultaneously. However strong your Macintosh may be, it can still perform only one calculation at a time. Why? 'Cause it's damned digital. But the time it loses in calculating one calculation at a time is covered up with the simple rapidness of the calculations it performs.)

To simulate the brain, our calculations in the real, digital world need to happen in parallel (because they need to be fast enough to match the freaking speed of the brain), and so we need to go around inventing supercomputers with millions of processors that can help us achieve this speed. The strongest supercomputer today resides in China, known as HPC 2010 China, which can already perform two quadrillion calculations in a second. That sort of a processing speed can process information twice as fast as the average human brain. This means that we definitely have the hardware to simulate a brain, and we just require the software to upload into our hardware. But that's the damn tough part.

What does our software really include? EVERYTHING. That's right, it isn't a misprint. We have to have the mathematical model for every neuron that is present in the mass of jelly. We need to have a complete and holistic understanding of the brain, especially and importantly in terms of the math behind it all. Once we have completely scanned and understood the (sadly) mathematical areas of the smelly jelly, then we can proceed to upload it on China's HPC 2010 and watch a computer, a literal machine, define intelligence all over again.

Wait up, hold on, wait a second. That's not it. There is, as usual, a problem.

Say that the scientists decide to scan my brain as a sample and upload it into the HPC2010. So, when my brain gets scanned and rigged into the computer, will I *myself* go into the computer? Or will the program on the code just be an exact imitation of me? Think about it. I mean, we don't even have the slightest idea about what really makes us *us,* so how can we be so sure that it will definitely be us who go into that damn computer? Can a computer really be a conscious human? Can a computer program really imitate us or *be* us? To that, my friends, there is no answer.

Or, there may as well be one in the next book I'm about to write.

Bibliography

1. " Is Time Travel Possible? - YouTube ." YouTube - Broadcast Yourself. . N.p., n.d. Web. 30 Jan. 2012. <http://www.youtube.com/watch?v=X02WMNoHSm8>.

2. Angier, Natalie. The canon: a whirligig tour of the beautiful basics of science. Boston: Houghton Mifflin Co., 2007. Print.

3. Capra, Fritjof. The Tao of physics: an exploration of the parallels between modern physics and eastern mysticism. Berkeley: Shambhala ;, 1975. Print.

4. Capra, Fritjof. The web of life: a new scientific understanding of living systems. New York: Anchor Books, 1996. Print.

5. Dawkins, Richard. The Oxford book of modern science writing. Oxford: Oxford University Press, 2008. Print.

6. Greene, B.. The fabric of the cosmos: space, time, and the texture of reality. New York: A.A. Knopf, 2004. Print.

7. HAWKING, STEPHEN. "STEPHEN HAWKING: How to build a time machine | Mail Online." Home | Mail Online. N.p., n.d. Web. 27 Apr. 2011. <http://www.dailymail.co.uk/home/moslive/article-1269288/STEPHEN-HAWKING-How-build-time-machine.html>.

8. Hawking, S. W.. A brief history of time: from the big bang to black holes. Toronto: Bantam Books, 1988. Print.

9. Hawking, S. W.. The theory of everything: the origin and fate of the universe. Beverly Hills, CA: New Millennium Press, 2002. Print.

10. "How Time Travel Works." How Stuff Works . Discover , n.d. Web. 20 Oct. 2011. <science.howstuffworks.com/science-vs-myth/everyday-myths/time-travel.html >.

11. Kaku, Michio. Hyperspace: a scientific odyssey through parallel universes, time warps, and the tenth dimension. New York: Oxford University Press, 1994. Print.

12. Kaku, Michio,. Parallel worlds: a journey through creation, higher dimensions, and the future of the cosmos. New York: Doubleday, 2005. Print.

13. Kaku, Michio. Physics of the impossible: a scientific exploration into the world of phasers, force fields, teleportation, and time travel. New York: Doubleday, 2008. Print.

14. Kaku, Michio, and Michio Kaku. Physics of the future: how science will shape human destiny and our daily lives by the year 2100. New York: Doubleday, 2011. Print.

15. Ramachandran, V. S., and Sandra Blakeslee. Phantoms in the brain: probing the mysteries of the human mind. New York: William Morrow, 1998. Print.

16. Ramachandran, V. S.. The tell-tale brain: a neuroscientist's quest for what makes us human. New York: W.W. Norton, 2011. Print.

17. "The Future of Physics: Scientific American." Science News, Articles and Information | Scientific American. N.p., n.d. Web. 22 Oct. 2011. <http://www.scientificamerican.com/article.cfm?id=the-future-of-physics>.

www.ingramcontent.com/pod-product-compliance
Lightning Source LLC
Chambersburg PA
CBHW030452220526
45464CB00006B/2497